GOODNIGHT PHARM

350 Brand
and
Generic Drug Names
with
Classifications

Tony Guerra, Pharm.D.

2017

Goodnight Pharm: 350 Brand and Generic Drug Names with Classifications

First Edition

ISBN: 9781387018116

To Mindy,

Brielle, Rianne, and Teagan

Table of Contents

PROLOGUE

The print, eBook, and audiobook versions of a book can have various different uses. The eBook matches the audiobook so you can follow along, but the print version, that's a different story. The print version is something you should use as an NCLEX®, NAPLEX®, or USMLE® Step-1 practice question journal. Pharmacology questions generally have a medication name in them, if so, you can add just the most important information.

For example, if a question outlines how furosemide, a loop diuretic causes hypokalemia, then turn to the furosemide page and write down "furosemide, hypokalemia." Taking the time to build your own journal from classes, test-prep books, and lectures will help you sort pharmacology information from many different information resources.

AUTHOR'S NOTE

In college and beyond, anxiety comes from looking at all you have to do in the future. Depression comes from regretting what you should have done in the past. Relaxation and easy learning comes from putting in your earphones, closing your eyes, and focusing on the present. Don't try to muscle these drug names into your memory, just...relax...listen...and you'll pick things up just as you would pick up the chorus in a song.

My pharmacology students work late night shifts or two jobs; try to balance work, school, and sometimes kids of their own – alone. A memorization-heavy class like pharmacology requires a lot of time they don't have. And studying for the boards, well, that's something even more challenging. When I ask my students what works, they give me three solutions:

1. Do a little bit each day

2. Use audio learning in the spaces of time they have

3. Designate the same time to do something each day

This book considers those three solutions.

1) You can listen to a chapter a day and be done in a week (Note: there is a bonus chapter 8 with newer drugs at the end)

2) This audiobook lets you listen while you are cooking, bathing, shaving, running, traveling, or commuting

3) As a parent of three five-year-olds, I know how difficult doing something at the same time each day is. But I recommend that you develop the habit of listening to an

audiobook before you go to sleep. It might not always be the same clock time, but this one habit will make many things better.

If you're good with drug names as you will be after listening to this book:

You're quicker to know what a patient or clinician is talking about. You're quicker to pick up on a concept in a test prep book.
You're quicker to answer board questions leaving more time for others.
You're quicker to get that stinking computer to get you the information you want.

This book will make you much quicker at many of the things that slow you down; leaving more time for things you might care about more than studying pharmacology like family, friends, and above all – sleep.

ACKNOWLEDGEMENTS

I could never have completed this book without the support and love of my wife Mindy, or the continued warmth and lessons I receive from my daughters Brielle, Rianne, and Teagan.

Introduction

I heard from my students that for pharmacology class, the audiobook *Memorizing Pharmacology: A Relaxed Approach* and learning 200 drugs was just the right number, but for their board exams, the NCLEX, NAPLEX, and USMLE Step 1 for example, they were worried they didn't know enough drug names. They heard horror stories of students getting drugs they'd never seen before and failing their exams.

This comprehensive 350 generic and brand name list with drug classes supplements the 200 drugs in the audiobook *Memorizing Pharmacology* with new mnemonics. While I'll write a stand-alone test prep book this year, I want to provide audio help for students taking their boards soon.

While this book builds on the 200 drugs from *Memorizing Pharmacology*, you can use it as a stand-alone book as well. If you haven't read or listened to *Memorizing Pharmacology*, here's what you need to know.

1. These 350 drugs are in a specific order and you should memorize them in this order. For example, the antacids come before the H2 blockers before the PPIs because that's the order of speed that they work in. The first generation antihistamines come before the second-generation antihistamines because you can look for an improvement, a lack of sedative adverse effects in the second generation. Within each class, I alphabetize the drugs to make it easy to recall.

2. Many generic drug names have prefixes and suffixes that have meaning. Many YouTube videos get this all wrong. My prefixes and suffixes come from the published lists from the

World Health Organization and United States Adopted Names Council; I didn't get cute and make up my own.

3. The ultimate takeaway from this book is looking at the drug name differently, using a different part of your brain than the linear logical to be creative and find meaning from words that don't give up their meaning easily.

I firmly believe you'll be more confident, better prepared and less anxious for your boards after listening to this book even if you don't write a single note down.

Pharmacology, I believe, will finally make sense and your life will be infinitely easier.

If I can improve in any way, please, please do take the time to let me know your needs by contacting me at aaguerra@dmacc.edu [[a-a-g-u-e-r-r-a at D-M-A-C-C dot e-d-u]] or on Facebook messenger.

CHAPTER 1 - GASTROINTESTINAL

I. Peptic Ulcer Disease

<u>Antacids</u>

1) Calcium carbonate (Tums, Children's Pepto) *(TUMS)*

cal'   = cal(orie)
ci   = (pronun)ci(ation)
um   = (g)um

car'   = car
bo   = bo(ne)
nate   = (in)nate

Students associate the brand name **Tums** with the word tummy to remember it's an antacid.

Calcium carbonate (Children's Pepto) and **bismuth subsalicylate (Adult Pepto-Bismol)** have different ingredients and shouldn't be interchanged. Remember, this is because the salicylate ingredient in the Adult Pepto can cause Reyes [[rise]] syndrome.

2) Magnesium Hydroxide (Milk of Magnesia)
(MILK of mag-KNEE-shuh)

mag   = mag(azine)
ne'   = (mo)ne(y)
si   = (enthu)si(astic)
um   = (g)um

hy   = hy(brid)
drox' = dr(ink)(b)ox
ide   = (sl)ide

Milk of Magnesia looks like dairy milk, which some still use to calm acidic stomachs. Put together the "milky texture" and the diarrhea of lactose intolerance to remember its laxative effect. What will happen if antacids don't work? Patients can choose between two other acid reducer classes: the histamine 2 receptor antagonists or the proton pump inhibitors.

Histamine-2 Receptor Antagonists (H₂RAs)

3) Famotidine (Pepcid) *(PEP-sid)*

fa	= (so)fa
mo'	= mo(at)
ti	= ti(c)
dine	= (bir)di(e)(tu)ne

Famotidine has the "-tidine," t-i-d-i-n-e stem indicating it's an H₂ blocker, but how do we remember what the brand name Pepcid is for? **Pepcid** contains "pep" from "peptic" and "pepsin" an enzyme related to digestion plus the "cid" from "acid." One student mentioned Pepsi, the soda, has a pH of 2.4, which is *very* acidic. She said it always gave her heartburn, so that's how she remembered **Pepcid**. Also, soda pop's "p-o-p" and **Pepcid's** "p-e-p" differ by only one letter.

4) Ranitidine (Zantac) *(ZAN-tack)*

ra	= (au)ra
ni'	= (k)ni(t)
ti	= ti(c)
dine	= (bir)di(e)(tu)ne

One student said **ranitidine's** "-tidine" stem looks like "to dine," the time patients might experience gastroesophageal reflux disease (GERD). The brand **Zantac** has a "Z" that looks like a "2" with "antac" after it, so it's an H *two* blocker, working as an _ant_-agonist to _ac_-id. Putting the 2 with "ant-ac" helps us remember Zantac is an H 2 blocker.

Proton Pump Inhibitors (PPIs)

With the proton pump inhibitors, **esomeprazole** (the (S) enantiomer of **omeprazole**) and **dexlansoprazole** (the (R)(+) enantiomer of **lansoprazole**) could be considered superior and segregated.

Nevertheless, I just put the PPIs in alphabetical order by generic name: **dexlansoprazole (Dexilant)**, **esomeprazole (Nexium)**, **omeprazole (Prilosec)**, **lansoprazole (Prevacid)**, **pantoprazole (Protonix)**, and **rabeprazole (AcipHex)**. Some of these newer brand names cleverly hint at function:

5) Dexlansoprazole (Dexilant) (DECKS-ih-lent)

The "prazole, p-r-a-z-o-l-e" stem reminds us each drug is a proton pump inhibitor. The brand Dexilant takes the "Dex" and "lan" from the generic name dexlansoprazole.

dex	= dex(terity)
lan	= lan(ce)
so'	= so(fa)
pra	= (su)pra
zole	= (fe)z(p)ole

6) **Esomeprazole (Nexium)** *(NECK-see-um)*

es'	= (m)es(s)
o	= "o"
me'	= me(n)
pra	= (su)pra
zole	= (fe)z(p)ole

The "prazole, p-r-a-z-o-l-e" ending again helps us remember it's a proton pump inhibitor, and the "e-s" means "S" for sinister or originating from the left. But how does the name Nexium relate to hyperacidic states? The manufacturer released **Nexium** after **Prilosec** as the "next" PPI drug.

7) Ome<u>prazole</u> (Prilosec) *(PRY-low-sec)*

o = "o"
me' = me(n)
pra = (su)pra
zole = (fe)z(p)ole

Just as with esomeprazole, the "prazole" ending in **omeprazole** indicates a proton pump inhibitor. We remember **Prilosec**'s relation to acid reduction by cutting the brand name into three parts.

The "Pr" (p-r) stands for hydrogen <u>pr</u>otons (protons, or hydrogen ions, are what make an acid acidic).
The "lo"(l-o) is for <u>lo</u>w, the opposite of high.
The "sec" (s-e-c) stands for proton <u>sec</u>retion.

Put those three parts together and you get "protons low secretion." Another way to remember Prilosec's function, and maybe a little less demanding, is to read "Pril oh sec" as "Pril-ZERO-sec" and remember it provides zero heartburn.

8. Lanso<u>prazole</u> (Prevacid) *(PREH-vuh-sid)*

lan = lan(ce)
so' = so(fa)
pra = (su)pra
zole = (fe)z(p)ole

Prevacid will "<u>prev</u>ent <u>acid</u>."

9. Panto<u>prazole</u> (Protonix) *(Pro-TAH-nicks)*

pan = pan
to' = to(e)
pra = (su)pra
zole = (fe)z(p)ole

Protonix "<u>nix</u>es <u>pro</u>tons."

10. Rabeprazole (AcipHex) *(ASS-eh-feks)*

ra   = (au)ra
be'   = be(d)
pra   = (su)pra
zole = (fe)z(p)ole

AcipHex combines "a-c-i" from acid, "pH, little 'p' capital 'H'" from the pH scale, and "ex" meaning to get rid of.

II. Diarrhea, constipation, and emesis

Under antidiarrheals, **diphenoxylate with atropine (Lomotil)** would alphabetically precede **loperamide**. I put it *after* **loperamide** because **diphenoxylate with atropine** represents an increase in the aggressiveness of treatment from OTC to prescription.

<u>Antidiarrheals</u>

11. Bismuth Subsalicylate (Pepto-Bismol) *(pep-TOE BIZ-mol)*

| bis' | = bis(quit) |
| muth | = (azi)muth |

sub	= sub(marine)
sa	= (vi)sa
li'	= li(ly)
cy	= cy(st)
late	= (p)late

Bismuth's stem is the "sal" in "sub*sal*icylate" and similar to **aspirin**, chemical name **acetyl*sal*icylic acid.** Both are dangerous to young children because of the risk of **Reye's [[rise]] syndrome**, a condition involving brain and liver damage that can occur in children with chicken pox or influenza who take **salicylates**. The "b" in **bismuth subsalicylate** reminds students of the black tongue and black stool that some patients experience as side effects. (Note: This discoloration is harmless.) The brand name **Pepto** looks like peptic, which has to do with digestion.

12) Loperamide (Imodium) *(eh-MOE-dee-um)*

lo	= (Co)lo(rado)
pe'	= pe(n)
ra	= (au)ra
mide	= (gu)m(sl)ide

The "lo" (l-o) for s<u>lo</u>w and "per" (p-e-r) for <u>per</u>istalsis is how one student remembered *loper*amide's function. **Imodium** is like the word "immobile" in that it slows down the bowel or immobilizes it.

13. Diphenoxylate / <u>atro</u>pine (Lomotil) (Low-MOE-till)

[[read forward slash as "with"]]

di = di(ce)
phe = phe(nomenon)
nox' = nox(ious)
yl = (s)yl(lable)
ate = (g)ate

a' = (m)a(t)
tro = (me)tro
pine = (shi)p(mar)ine

The brand name **Lomotil** spells out "<u>lo</u>w <u>moti</u>lity" for slowing down diarrhea. The **atropine** is there to prevent someone from crushing the **diphenoxylate** and injecting it illicitly.

<u>Constipation – Stool softener</u>

14) Docusate Sodium (Colace) *(CO-lace)*

do' = do(t)
cu   = cu(be)
sate = (pul)sate

so' = so(fa)
di   = (bir)di(e)
um   = (g)um

Docusate sodium softens the stool and patients use it with opioids like **morphine. Docusate** and "penetrate" rhyme to remind us that **docusate sodium** works by helping water *penetrate* into the bowel. The brand name **Colace** alludes to improving the <u>col</u>on's <u>pace</u>. Here we can use the c-o-l from "colon's" and a-c-e from "pace" to construct the brand name, Colace.

Constipation – Osmotic

15. Polyethylene Glycol (MiraLax) *(MEER-uh-lacks)*

po	= po(t)
ly	= (li)ly
eth'	= (m)eth(od)
yl	= (s)yl(lable)
ene	= (sc)ene
gly'	= gly(cemic)
col	= col(lar)

I remember **polyethylene glycol's** function because I have triplets, "**poly,**" call "**c-o-l**" me Daaad! Daaaaad! Dad! Daddy! *just* as I'm about to sit down to dinner" Some remember **MiraLax** is the Miracle Laxative because it's a miracle how good you feel after taking it. **MiraLax** has a "by prescription only cousin." **Go-Lytely** is a 4-liter plastic bottle of **polyethylene glycol** used for colonoscopy examination preparation. There is nothing "lightly" about its laxative effect, so you will want to make sure there is a bathroom nearby.

Constipation – Miscellaneous

16) Lubiprostone (Amitiza) (ah-meh-TEA-zuh)

lu   = lu(cid)
bi   = bi(t)
pro' = pro(m)
stone = stone

I put **lubiprostone** after **docusate sodium** and **polyethylene glycol** because it's a prescription item and would represent an escalation in the aggressiveness of treatment for constipation. While the "-prost-" stem indicates prostaglandin, it doesn't really help with immediate therapeutic recognition. With the laxative **lubiprostone**, I would think of using lube to propel a stone out of the body.

Antiemetic – Serotonin 5-HT$_3$ receptor antagonist

17. Ondansetron (Zofran, Zofran ODT) *(ZO-fran)*

on	= on
dan'	= dan(ce)
se	= se(t)
tron	= (elec)tron

The "setron" s-e-t-r-o-n suffix will help you remember
ondansetron is a serotonin 5-HT$_3$ receptor antagonist for
preventing emesis. If you are good at word scrambles,
ondansetron has every letter but the "i" in serotonin, a
neurotransmitter, the majority of which is located in the GI
tract. The "O-D-T" in Zofran O-D-T stands for orally
disintegrating tablet. It's a useful dosage form because it
dissolves on top of the tongue without additional liquid.

<u>Antiemetic – Phenothiazine</u>

18) Prochlorperazine (Compazine) (COM-puh-zeen)

pro   = pro(ton)
chlor   = chlor(ine)
pe'   = pe(n)
ra   = (au)ra
zine   = (maga)zine

Prochlorperazine and **promethazine (Phenergan)** share the same first three letters, "p-r-o," and last five letters, "a-z-i-n-e." While this isn't a proper stem, both are phenothi<u>azine</u>s. By memorizing these two drugs in alphabetical order, you can use this similarity to recognize their comparable antiemetic function. Both happen to be available in suppository formulations as well.

19) Promethazine (Phenergan) *(FEN-er-gan)*

pro   = pro(ton)
meth' = meth(od)
a   = (comm)a
zine   = (maga)zine

Promethazine is technically an antihistamine manufacturers sometimes combine in liquid form with codeine.
Promethazine also reduces nausea. In addition to oral, IM, and IV forms, **promethazine** comes in a rectal suppository if a patient can't take anything by mouth (po) [[Read as p-o]].

III. Autoimmune disorders

<u>Ulcerative colitis</u>

20. Infliximab (Remicade) *(REM-eh-cade)*

in   = in
flix'   = fl(our)(m)ix
I   = i(t)
mab = m(at)(c)ab

Is a biologic agent, a genetically engineered protein.

Break up **infliximab's** generic name as (i-n-f) + (l-i) + (x-i) + mab (m-a-b).

The "i-n-f" is a prefix that separates it from other similar drugs.

The "l-i" stands for immunomodulator (the target).

The "x-i" stands for chimeric, the source. For example, combining genetic material from a mouse, with genetic material from a human.

The "x" might also refer to the Greek letter "chi," which looks like an x.
The "mab" (m-a-b) stands for monoclonal antibody.

Conditions like ulcerative colitis can go into remission so you'll want to think of **Remicade** as a "remission aide," where the letters r-e-m-i from "remission" and a-d-e from "aide" are prominent in the brand name; **Remicade**.

CHAPTER 2 - MUSCULOSKELETAL

I. NSAIDs [[en-saids]] and pain

OTC Analgesics – NSAIDs

21) Aspirin (Ecotrin) *(ECK-oh-trin)*

as'   = as(p)
pi   = pi(n)
rin   = rin(se)

The acronym for **aspirin**, "A-S-A," comes from the chemical name: Acetylsalicylic Acid, taking the "A" from acetyl, "S" from salicylic, and "A" from acid.

The brand **Ecotrin** is enteric-coated aspirin, taking the "e" from "enteric," the c-o and t from "coated," and the r-i-n from "aspirin" to make **Ecotrin**.

22) Ibuprofen (Advil, and Motrin) *(ADD-vil, MO-trin)*

i   = "i"
bu   = bu(reau)
pro'   = pro(ton)
fen   = fen

Many students try to say that **ibuprofen** and **naproxen** both end with "e-n" and that's a good way to remember their relationship as two NSAIDs.

However, many drugs end with "e-n," so that won't help in a large multiple-choice exam.

A better mnemonic is to notice that "profen, p-r-o-f-e-n" from **ibuprofen** is a recognized stem and differs from "proxen, p-r-o-x-e-n" in **naproxen** by only one letter.

23) Naproxen (Aleve) *(uh-LEAVE)*

na   = (bana)na, fexofenadine
prox' = pr(oton)(b)ox, naproxen
en   = (p)en, (m)en, influenza vaccine

While **naproxen** has no formal stem, a student came up with a mnemonic for the brand name: **Aleve will** <u>allev</u>iate pain from strains and sprains.

OTC Analgesic – Non-narcotic

24. Acetaminophen (Tylenol) *(TIE-len-all)*

a	= (comm)a
ce	= ce(iling)
ta	= (da)ta
mi'	= mi(nt)
no	= (can)no(n)
phen	= (hy)phen

The brand **Tylenol**, generic **acetaminophen**, and acronym **A-PAP** all come from the chemical name:
N-acetyl-para-amino-phenol

From which we take the t-y-l from "acetyl" and the e-n-o-l ending from "phenol" for the brand name **Tyl-enol**

We take most of the words "acetyl," "amino" and "phenol" for the generic name, **Acetaminophen**.

And we take the four first letters for the acronym, A-P-A-P, or A-PAP.

OTC Migraine – NSAID / Non-narcotic analgesic

25. Aspirin with Acetaminophen and Caffeine (Excedrin Migraine) *(ecks-SAID-rin)*

as'	= as(p)
pi	= pi(n)
rin	= rin(se)

a	= (comm)a
ce	= ce(iling)
ta	= (da)ta
mi'	= mi(nt)
no	= (can)no(n)
phen	= (hy)phen

caf'	= caf(e)
feine	= fe(et)(magaz)ine

Most students remember **Excedrin Migraine** by the rationale for the combination of **aspirin with acetaminophen and caffeine** that is inflammation, analgesia, and vasoconstriction respectively.

RX Migraine – Narcotic and Non-narcotic analgesic

26. Butalbital / Acetaminophen / Caffeine (Fioricet) *(Fee-OR-ih-set)*

bu	= bu(reau)
tal'	= tal(l)
bi	= bi(t)
tal	= tal(l)
a	= (comm)a
ce	= ce(iling)
ta	= (da)ta
mi'	= mi(nt)
no	= (can)no(n)
phen	= (hy)phen
caf'	= caf(e)
feine	= fe(et)(magaz)ine

Adding **butalbital**, a barbiturate, to **acetaminophen** and **caffeine** in **Fioricet** provides an escalation from the OTC **aspirin/acetaminophen/caffeine** combination in **Excedrin Migraine**. I pair those two drugs in my mind to help remember their therapeutic function.

RX Analgesics – NSAIDs

For the NSAIDs I simply alphabetized four additions around **meloxicam**.

27. Diclofenac sodium extended release (Voltaren XR) *(vol-TARE-in)*

di   = di(ce)
clo'   = clo(ver)
fe   = fe(n)
nac   = (k)nac(k)

 is a generic name derived from its chemical structure 2-[(2,6-dichlorophenyl)amino] benzeneacetic acid with a change from "p-h" to "f" in the middle. [[Note: Doing a chemical name pronunciation is actually kind of fun, here are the two for that chemical compound:

Two Two six die klor' oh fen' ill uh me' no Ben' zeen uh see' tic a cid'

Two, Two, Six
Di(e), Chlor(ine)', "o"
(hy)phen', (vin)yl
(aur)a, (a)mi', no(se)
(Mercedes) benz', (sc)ene
(aur)a, ce(iling)', tic
acid]]

28. Etodolac (Lodine) *(LOW-deen)*

e   = (n)e(t)
to'   = to(e)
do   = (i)do(l)
lac   = (li)lac

Shares the same generic stem, "a-c, ac" as diclofenac.

29. Indomethacin (Indocin) *(IN-doe-sin)*

in   = in
do   = do(ugh)
meth'   = meth(od)
a   = (comm)a
cin   = cin(der)
(Indocin)

The manufacturers of **indomethacin** removed the middle "metha" to get the brand name **Indocin**.

30. Meloxicam (Mobic) *(MO-bik)*

me   = me(n)
lox'   = lox
i   = i(t)
cam   = cam(p)

The "-icam,"-i-c-a-m suffix in the generic name **meloxicam** lets you know it's an NSAID. A student used the "bic" in **Mobic** to remember it treats "big" swelling.

31. Nabumetone (Relafen) *(RELL-uh-fin)*

na   = (bana)na
bu'   = bu(reau)
me   = me(n)
tone   = tone

Nabume-tone tones down pain. Nabumetone's brand name **Relafen** sounds a little like getting pain relief with an "en-said."

RX Analgesics – NSAIDs – COX-2 inhibitor [[cocks two]]

32. Cele<u>coxib</u> (Celebrex) *(SELL-eh-breks)*

ce	= ce(nt)
le	= le(tter)
cox'	= cox(swain)
ib	= (r)ib

The "–coxib, c-o-x-i-b" suffix lets you know celecoxib is a selective COX-2 inhibitor. This contrasts regular NSAIDs like **ibuprofen** and **naproxen** that inhibit both COX-1 and COX-2, causing the stomach distress so commonly caused by drugs in the NSAID class. **Celecoxib** causes less G-I irritation due to its lack of COX-1 inhibition.

The commercials for brand **Celebrex** talk about <u>celebra</u>ting relief from inflammatory conditions.

II. Opioids and narcotics

Opioid analgesics – Schedule II

33. Morphine (Kadian or M-S Contin) *(KAY-dee-en, EM-ES KON-tin)*

mor' = mor(e)
phine = (So)phi(a)(du)ne

The generic name **morphine** comes from the ancient Greek god of dreams, Morpheus.

The brand **Kadian** might come from cir<u>cadian</u> (the twenty-four-hour cycle) because **Kadian** is an extended-release morphine formulation.

M-S Contin stands for <u>m</u>orphine <u>s</u>ulfate **contin**uous release, referencing the M-S from the beginnings of "morphine sulfate," and the "Contin" from the start of "continuous;" **M-S Contin**.

34. **Fentanyl (Duragesic or Sublimaze)** *(dur-uh-GEE-zic, SUB-leh-maze)*

fen' = fen
ta = (da)ta
nyl = (vi)nyl

Fentanyl is troubling because it's dosed in micrograms, not milligrams. When it was time to give our three-month-old preemie **fentanyl** after her pyloric valve stenosis surgery in the Pediatric Intensive Care Unit (PICU) [[pick-you]], I made sure to check the calculated dose.

Duragesic is a long <u>dura</u>tion anal<u>gesic</u> and comes in a patch that provides relief for 72 hours; our mnemonic for Duragesic takes the "dura" from a long "duration" and the "gesic" from the ending of "analgesic."

Sublimaze is an injectable form of **fentanyl**.

In combination medications, I put the shared generic drug first, followed by the additional drug. After **hydrocodone / acetaminophen (Vicodin)**, I followed with the **hydrocodone / chlorpheniramine (Tussionex)** and **hydrocodone / ibuprofen (Vicoprofen)**.

35. Hydrocodone with Acetaminophen (Vicodin) *(VY-co-din)*

hy = hy(brid)
dro = dro(ne)
co' = co(ne)
done = (con)done

a = (comm)a
ce = ce(iling)
ta = (da)ta
mi' = mi(nt)
no = (can)no(n)
phen = (hy)phen

Hydrocodone and **oxycodone** differ slightly in chemical structure. **Oxycodone** has one more oxygen atom than **hydrocodone**.

36. Hydrocodone / Chlorpheniramine (Tussionex) *(TUSS-ee-uh-necks)*

hy = hy(brid)
dro = dro(ne)
co' = co(ne)
done = (con)done

chlor = chlor(ine)
phe = phe(nomenon)
ni' = ni(ght)
ra = (au)ra
mine = (hista)mine

Tussionex is a pineapple-flavored liquid anti<u>tussi</u>ve, hence the brand name.

37. Hydrocodone / Ibuprofen (Vicoprofen) *(vie-ko-PRO-fin)*

hy	= hy(brid)
dro	= dro(ne)
co'	= co(ne)
done	= (con)done
I	= "i"
bu	= bu(reau)
pro	= pro(ton)
fen	= fen

Vicoprofen is like **<u>Vicodin</u>**, but with **ibu<u>profen</u>** instead of **acetaminophen**. Therefore, the manufacturer replaced the "i-b-u" of **ibuprofen** with "V-i-c-o" of **Vicodin**.

38. Methadone (Dolophine) *(DOE-luh-feen)*

meth'	= meth(od)
a	= (comm)a
done	= (con)done

I put **methadone** in alphabetically although its primary purpose is to help patients addicted to opiates get off narcotics, not provide pain relief.

39. Oxycodone (OxyIR, Oxycontin) [[ock-see I-R]] *(OX-ee-eye-are, OX-ee-con-tin)*

ox	= ox
y	= (countr)y
co'	= co(ne)
done	= (con)done

Oxycodone by itself comes in an immediate release form, **OxyIR**, and an extended release form, **OxyContin**. The "c-o-n-t-i-n" means <u>contin</u>uous release. Many people mispronounce this as oxy-cotton, using a word they are familiar with.

40. Oxycodone with Acetaminophen (Percocet) *(PER-coe-set)*

ox	= ox
y	= (countr)y
co'	= co(ne)
done	= (con)done

a	= (comm)a
ce	= ce(iling)
ta	= (da)ta
mi'	= mi(nt)
no	= (can)no(n)
phen	= (hy)phen

The "cet" (c-e-t) in **Percocet** comes from one British pronunciation of a<u>cet</u>aminophen [[a set uh minnow fin]]. Some students use that "codone" and "codeine" look a little alike to remember the similarity, but this is not a true stem and **Percocet** doesn't contain codeine. Drug companies add **acetaminophen** as a mild analgesic.

CHAPTER 2 - MUSCULOSKELETAL

Opioid analgesics - Schedule III

41. Acetaminophen with Codeine (Tylenol with Codeine or Tylenol number 3) *(TIE-len-all with CO-dean, TIE-len-all NUMber three)*

a   = (comm)a
ce   = ce(iling)
ta   = (da)ta
mi'   = mi(nt)
no   = (can)no(n)
phen   = (hy)phen

co'   = co(ne)
deine   = de(er)

I am not sure why, but when you say the generic name of **Vicodin**, you say "**hydrocodone** *with* **acetaminophen**," but when you talk about **codeine,** the order is reversed. It's phrased as "**acetaminophen** *with* **codeine**"

Students seem to remember both because of the reverse and opposite order. The "#3" [[read # as "number"]] in **Tylenol #3** refers to the amount of codeine in combination. For example:

Tylenol #2 has 15 mg [[mg read as milligrams]] codeine with 300 mg acetaminophen

Tylenol #3 has 30 mg codeine with 300 mg acetaminophen

Tylenol #4 has 60 mg codeine with 300 mg acetaminophen

27

<u>Mixed-opioid receptor analgesic – Schedule IV</u>

42. Tram<u>adol </u>(Ultram) *(ULL-tram)*

tra'	= tra(ck)
ma	= (paja)ma
dol	= dol(l)

Tramadol only weakly affects opioid receptors. For this reason, the D-E-A did not classify **tramadol** as a controlled substance until 2014. The "adol, a-d-o-l" stem indicates that it's a mixed opioid analgesic.

Many students think of "tram wreck" and "train wreck" as a way to remember that **Ul*tram*** is used for pain.

43. Tram<u>adol</u> / Acetaminophen (Ultracet) *(Ul-truh-set)*

tra'	= tra(ck)
ma	= (paja)ma
dol	= dol(l)

a	= (comm)a
ce	= ce(iling)
ta	= (da)ta
mi'	= mi(nt)
no	= (can)no(n)
phen	= (hy)phen

Tramadol with **acetaminophen (Ultracet)** follows **tramadol** by itself. The manufacturer took the brand name **Ultram** (for **tramadol** alone), dropped the "m" and added "acet" from **acetaminophen**.

Opioid antagonist

44. Naloxone (Narcan) *(NAR-can)*

na   = (bana)na
lox'   = lox
one   = (c)one

The "nal, n-a-l" stem in **naloxone** indicates it's an opioid receptor antagonist, but the brand name **Narcan** with its elements "Narc" and "an," also hints at a narcotic antagonist.

Naloxone is an opioid receptor antagonist used in opioid overdose situations. Often it's in the L-E-A-N, lean, acronym of emergency medicines lidocaine, epinephrine, atropine, and naloxone.

45. Buprenorphine / Naloxone (Suboxone) *(sue-BOX-own)*

bu   = bu(reau)
pre   = pre(dator)
nor'   = nor
phine = (So)phi(a)(du)ne

na   = (bana)na
lox'   = lox
one   = (c)one
(sue-BOX-own)

[[Schedule C-III]]

Patients use **buprenorphine / naloxone** like **methadone** to help detox from opiate addiction. The **naloxone** is there to keep patients from crushing the drug and injecting it.

III. Headaches and migraine

<u>5-HT$_1$ receptor agonist</u>

46) Ele<u>trip</u>tan (Relpax) *(rel-PACKS)*

e	= (n)e(t)
le	= le(tter)
trip'	= trip
tan	= (sul)tan

Remember **eletriptan** from its suffix "triptan, t-r-i-p-t-a-n" so you can recognize the many other medications in the triptan class.

The brand name **Relpax** combines "Rel" for "relief" and "pax," the Latin word for "peace." I often confused whether **Relpax** was an agonist or antagonist, but use the "agony" of a migraine to remember triptans as agonists.

47. Suma<u>trip</u>tan (Imitrex) [[im like him, not im like I'm]] *(IM-eh-treks)*

su	= su(e)
ma	= (paja)ma
trip'	= trip
tan	= (sul)tan

Remember **sumatriptan** from its suffix "–triptan." Students say it "trips up" a headache. The injectable form is for patients who have such a severe migraine that they can't take anything by mouth.

IV. DMARDs [[dee-mards]] and rheumatoid arthritis

48. Metho<u>trexate</u> (Rheumatrex) *(ROOM-uh-treks)*

meth = meth(od)
o   = "o"
trex' = tre(mor)
ate = (g)ate

The "trexate, t-r-e-x-a-t-e" stem helps remind you this is a DMARD. One student came up with "Meth o T-Rex ate the rheumatic inflammate;" **Methotrexate**.

The "rheuma" in the brand name **Rheumatrex** reminds you it relieves rheumatoid arthritis.

49. Aba<u>tacept</u> (Orencia) *(or-EN-see-uh)*

a   = (comm)a
ba   = ba(t)
ta' = (da)ta
cept = (ac)cept

Abatacept has a complex stem. The "-ta-", t-a, infix in **abatacept** means it's going after T-cell receptors and the suffix "–cept, c-e-p-t" means that it's a re<u>cept</u>or molecule, either native or modified.

50. Eta<u>nercept</u> (Enbrel) *(EN-brell)*

e   = (en)e(my)
ta' = ta(b)
ner = ner(d)
cept = (ac)cept

Learn both **etanercept** and **abatacept** from the suffix "–cept" (c-e-p-t), with the added sub-stem "-nercept" or "-tacept" respectively.

The "-ner-" (n-e-r) infix in **etanercept** points out that it goes after tumor <u>ner</u>crosis factor receptors, where we're noting the letters n-e-r from "necrosis."

V. Osteoporosis

<u>Bisphosphonates</u>

51. Alen<u>dronate</u> (Fosamax) *(FA-seh-max)*

a   = (comm)a
len' = len(s)
dro   = dro(ne)
nate = (in)nate

Memorize **alendronate** as a calcium metabolism regulator from its "dronate, d-r-o-n-a-t-e" stem. Students like to remember that "drone" rhymes with bone.

Many students have said that the brand name **Fosamax** looks like a "fossil."

52. Iban<u>dronate</u> (Boniva) *(bo-KNEE-vuh)*

i   = "i"
ban' = ban(d)
dro   =dro(ne)
nate = (in)nate

Again, the "dronate" suffix should be your key to the drug class, but having the first three letters of bone in the brand name **Boniva** helps. The generic **ibandronate** contains all the letters in **Boniva** except the "v."

53. Rise<u>dronate</u> (Actonel) *(AK-tuh-nell)*

ri   = ri(nse)
se'   = se(t)
dro   = dro(ne)
nate = (in)nate

is another bisphosphonate that you can recognize from the "-dronate" stem.

VI. Selective estrogen receptor modulator (SERM)

54. Raloxifene (Evista) *(ee-VIS-tuh)*

ra	= (au)ra
lox'	= lox
i	= i(t)
fene	= fe(et)(du)ne

We classify the <u>s</u>elective <u>e</u>strogen <u>r</u>eceptor <u>m</u>odulator (SERM) **ralox<u>ifene</u> (Evista)** as an antiestrogen by its "–ifene" stem.

VII. Muscle relaxants

There are many muscle relaxants and I have simply alphabetized them by generic name: **baclofen (Lioresal)**, **carisoprodol (Soma)**, **metaxalone (Skelaxin)**, **methocarbamol (Robaxin)**, and **tizanidine (Zanaflex)**.

55. Baclofen (Lioresal) *(Lee-OR-uh-sahl)*

ba'   = ba(t)
clo   = clo(ver)
fen   = fen

Turn the "l-o, lo" and "b-a-c-, bac" from baclofen to remember it can be used for "low back" pain.

56. Carisoprodol (Soma) *(SEW-muh)*

ca   = (or)ca
ri   = ri(nse)
so   = so(n)
pro' = pro(ton)
dol   = dol(l)

Some of the brand names hint at muscle relaxation, e.g., **Soma** sounds like <u>som</u>nolence which means sleepiness. If you are a literature nerd, this drug was a hallucinogen in Aldous Huxley's 1932 book, *Brave New World*.

57. Cyclobenzaprine (Flexeril) *(FLEX-er-ill)*

cy   = cy(cle)
clo   = clo(ver)
ben'   = ben(d)
za   = (pla)za
prine = pr(oton)

Cyclobenzaprine helps you get bending again, using the 'b-e-n-z' in the center of the generic name. The brand *Flexeril* improves *flex*ibility.

58. Di<u>azepam</u> (Valium) *(VAL-e-um)*

di   = di(ce)
a'   = (m)a(t)
ze   = ze(d)
pam   = pam(phlet)

Diazepam and benzo**diazep**ine, diazepam's drug class, have similar letters. Note the v-a-l in <u>Val</u>erian root, which is an herbal remedy for anxiety has the same three initial letters as **Valium**. Some think of **Valium** as relaxing both anxiety and muscles in the same way.

59. Metaxalone (Skelaxin) *(skeh-LAX-in)*

me   = me(n)
tax'   = tax
a   = (comm)a
lone   = lone

Skelaxin alludes to "<u>skel</u>etal rel<u>axin</u>'."

60. Methocarbamol (Robaxin) *(row-BACKS-in)*

meth   = meth(od)
o   = "o"
car'   = car
ba   = tu(ba)
mol   = mol(lusk)

Robaxin and rel<u>axin</u>' go together.

61. Tizanidine (Zanaflex) *(ZA-nuh-flecks)*

ti   = ti(c)
za'   = (pla)za
ni   = (k)ni(t)
dine   = (bir)di(e)(tu)ne

Alternate pronunciation:

ti = ti(e)
za' = (pla)za
ni = (k)ni(t)
dine = (bir)di(e)(tu)ne

Zanaflex ends with "flex" for increased fle<u>x</u>ibility.

VIII. Gout

62. Colchicine (Colcrys) *(COAL-Chris)*

col'   = col(lar)
chi   = chi(n)
cine   = (leu)cine

Colchicine relieves an acute gouty attack, so I put that before the uric acid reducers that treat chronic increased uric acid. The "crys" in **Colcrys** sounds like the painful gouty crystals that often form in the big toe. Google "gouty crystal" images and you'll see that they look like needles.

Uric acid reducers

63. Allopurinol (Zyloprim) *(ZY-low-prim)*

al   = (s)al(iva)
lo   = (Co)lo(rado)
pur'   = pur(e)
i   = i(t)
nol   = (etha)nol

Within **allopurinol**, you can see "uri, u-r-i," which corresponds to the uric acid the medication reduces. You can also remember this is an anti-arthritic by thinking of the joints as becoming "all-pure-and-all," mimicking the generic name, allopurinol.

64. Febuxostat (Uloric) *(YOU-lore-ick)*

fe   = fe(n)
bux' = bux(om)
o   = "o"
stat = (photo)stat

The "xostat, x-o-s-t-a-t" stem in **febuxostat** indicates a xanthine oxidase inhibitor that prevents uric acid from forming. The "x" and "o" in the stem included in the generic name match these first letters in xanthine oxidase. The brand name **Uloric** looks like "U" "lower" "uric acid;" **Uloric**.

CHAPTER 3 - RESPIRATORY

I. Antihistamines and decongestants

Antihistamine – 1st-generation

65. Diphenhydramine (Benadryl) *(BEN-uh-drill)*

di = di(ce)
phen = (hy)phen
hy' = hy(brid)
dra = (hy)dra
mine = (hista)mine

Many students on YouTube videos and Quizlet notecards mistake **diphenhydramine's** "i-n-e" as a stem because many antihistamines end in "ine." [[pronounced "een"]]

However, **morphine**, an opioid, ends in "een" and roughly 20% of *all* generic names do. There is not really a good stem for some generic antihistamine names because of this.

You can remember the brand name by recognizing **Benadryl** benefits you by drying up your runny nose; taking the "ben" from "benefits" and the "dry" from "drying."

Some students also associated the capital "B" in **Benadryl** with the BBB, the Blood Brain Barrier, which **Benadryl** *can* pass through.

Manufacturers use **diphenhydramine** as the "P-M" in many sleep aids, so associating the "B," "e," and "d" in **Benadryl** with bedtime also makes sense.

66. Hydroxyzine (Atarax) *(AT-uh-raks)*

hy	= hy(brid)
drox'	= dr(ink)(b)ox
y	= (t)y(pical)
zine	= (maga)zine

I added the first-generation antihistamine **hydroxyzine** alphabetically after **diphenhydramine.** The "x-y-z" inside **hydro<u>xyz</u>ine** matches the brand name of a third-generation antihistamine **<u>Xyz</u>al**. [[zie-zal]]

OTC Antihistamine – 2nd-generation

67. Cetirizine (Zyrtec) *(ZEER-teck)*

ce   = ce(nt)
ti'   = ti(c)
ri   = ri(nse)
zine = (maga)zine

I pronounce the "t-i-r" in **cetirizine** as tear, like teardrop, and I think of **cetirizine** protecting me from tearing, from eyes affected by allergies.

68. Loratadine (Claritin) *(CLAIR-ih-tin)*

lo   = (Co)lo(rado)
ra'   = ra(ttle)
ta   = (da)ta
dine = (bir)di(e)(tu)ne

The "–atadine, a-t-a-d-i-n-e" stem (used to be "–tadine" t-a-d-i-n-e) helps distinguish **loratadine** from the "–tidine" (t-i-d-i-n-e) stem in the H_2 receptor antagonists **famotidine** and **ranitidine.**

The **Claritin** "clear" commercials resonate with the brand name drug's function of *clearing* one's head from allergies or *clear* eyes relieved from allergy.

OTC Antihistamine – 3nd generation

69. Fexofenadine (Allegra) *(uh-LEH-gruh)*

fex = (ponti)fex
o = "o"
fe' = fe(n)
na = (bana)na
dine = (bir)di(e)(tu)ne

The third-generation antihistamine **fexofenadine** is a safe active metabolite of **terfenadine (Seldane),** a drug the manufacturer removed from the market because of cardiac side effects.

70. Levocetirizine (Xyzal) *(ZIE-zuhl)*

le = le(tter)
vo = vo(te)
ce = ce(nt)
ti' = ti(c)
ri = ri(nse)
zine = (maga)zine

Is not the active metabolite, but rather the left enantiomer of **cetirizine (Zyrtec).** The commericials for the Xyzal owl mascot resonate as my kids have three stuffed Xyzal owl animals that talk and espouse the benefits of the drug.

OTC Antihistamine – Eye Drops

71. Olopatadine (brands Patanol, Pataday) *(PAH-tuh-nohl, PAH-tuh-day)*

o   = "o"
lo   = (Co)lo(rado)
pa'   = pa(th)
ta   = (da)ta
dine   = (bir)di(e)(tu)ne

From oral antihistamines, I moved to ophthalmic (eye) and nasal forms. **Olopatadine (Patanol, Pataday)** shares the "–atadine" antihistamine stem of **loratadine**. The two "o's" at the beginning of the generic **olopatadine** look like eyes.

<u>Antihistamine – Nasal Spray</u>

72. Aze<u>lastine</u> (Astelin) *(ah-STEH-Lynn)*

a = (m)a(t)
ze' = ze(d)
la = la(b)
stine = (cele)stine

Is a medicine you <u>stick</u> <u>in</u> your nose.

OTC Antihistamine – 2nd generation / Decongestant

73. Loratadine with Pseudoephedrine (Claritin-D) *(CLAIR-ih-tin dee)*

lo   = (Co)lo(rado)
ra'   = ra(ttle)
ta   = (da)ta
dine   = (bir)di(e)(tu)ne

Because **pseudoephedrine** has 15 letters, manufacturers abbreviate it as "hyphen D" for decongestant. Therefore, adding **loratadine**, an antihistamine, to **pseudoephedrine** a decongestant, helps with both runny and stuffy noses.

BTC/OTC Decongestants

74. Pseudoephedrine (Sudafed) *(Sue-duh-FED)*

pseu	= pseu(do)
do	= do(ugh)
e	= (n)e(t)
phe'	= phe(nomenon)
drine	= (alexan)drine

If you take out the second "e" and drop the "rine" (r-i-n-e) from **pseudoephedrine**, you get the pronunciation of **Sudafed**. One student said she was p-h-e-d up, "phed up," with being congested, and that's how she remembered it.

75) Phenylephrine (NeoSynephrine) *(KNEE-oh-sih-NEF-rin)*

phen	= (hy)phen
yl	= (s)yl(lable)
eph'	= (z)eph(yr)
rine	= (doct)rine

Phenylephrine sounds a bit like **pseudoephedrine** – they are both decongestants. Patients recognize *pseudoephedrine* by the "hyphen D" and *phenylephrine* by the "P-E" abbreviation in many cold preparations.

Phenylephrine is not as strong as **pseudoephedrine** and that's one reason it's available OTC, while **pseudoephedrine** is not.

76) Oxymetazoline (Afrin) *(AF-rin)*

ox   = ox
y   = (countr)y
me   = me(n)
ta'   = ta(b)
zo   = zo(ne)
line   = line(n) [Note: not (fe)line or (sa)line]

Oxymetazoline takes us from physically behind-the-counter (B-T-C) oral **pseudoephedrine** to *over*-the-counter nasal or oral **phenylephrine** to the *intranasal* decongestant **oxymetazoline**.

The "**Afrin**" brand name sounds a little like the "ephrine" that forms the ending of **phenyl*ephrine*** so you can relate the two as decongestants.

II. Allergic rhinitis steroid, antitussives, and expectorants

<u>Allergic rhinitis steroid</u>

77. Mometasone nasal inhaler (Nasonex) *(NAY-zuh-necks)*

mo   = mo(at)
me' = me(n)
ta   = (da)ta
sone = so(fa)(du)ne

Prescription drugs sometimes transition to over-the-counter (OTC), so I just put **mometasone** alphabetically before another allergic rhinitis steroid **triamcinolone**.

78. Triamcinolone (Nasacort Allergy twenty-four hour)
(NAY-zuh-cort)

tri = tri(angle)
am   = (l)am(p)
ci' = (s)ci(ntillate)
no   = (can)no(n)
lone = lone

There is no recognized stem in **triamcinolone**, but often "o-n-e," pronounced like I "own" something, matches the "o-n-e" at the end of **testosterone**, a more familiar steroid.

The brand name **Nasacort** Allergy twenty-four hour reads like a story: "Nasa" for nose, "cort" for corticosteroid, "Allergy" for allergic rhinitis, and "twenty four hour" for how long it works; Nasacort Allergy twenty four-hour.

OTC Antitussive / Expectorant

79. Guaifenesin with Dextromethorphan

(Mucinex DM and Robitussin DM) *(MEW-sin-ex dee-em, row-beh-TUSS-in dee-em)*

guai	= gua(va)"i"
fe'	= fe(n)
ne	= ne(t)
sin	= (ba)sin

dex'	= dex(terity)
tro	= (me)tro
meth	= meth(od)
or'	= or
phan	= (or)phan

Guaifenesin, pronounced as if you put a "g" in front of "why," is an expectorant, somethat that expels sputum from the chest. You can see "p-e-c-t-o-r, pector" in the word expectorant to remind you of pectoralis or chest muscles. Students remember this because Mr. Mucus from the **Mucinex**-brand commercials is green, and green also starts with a "g."

Robitussin "robs" your cough and "tussin" resembles "tussive." Anti*tussives* are anti-cough medicines.

RX Antitussive / Expectorant

80. Guaifenesin with Codeine (Cheratussin A-C) *(CHAIR-uh-tuss-in AY-see)*

guai = gua(va)"i"
fe'   = fe(n)
ne   = ne(t)
sin   = (ba)sin

Sometimes **dextromethorphan (DM)** isn't enough and the patient needs **codeine** to suppress a cough. Most students know **codeine**, but cough and **codeine** both start with "c-o" and that seems to help.

The "chera" in the brand **Cheratussin** comes from the product's cherry flavoring. Some are not sure what the "AC" means; some students think "anti-cough" although it's probably "and codeine."

RX Antitussive

81. Benzonatate (Tessalon Perles) *(TESS-ah-lahn pearls)*

ben   = ben(d)
zo'   = zo(mbie)
na   = (bana)na
tate   = (es)tate

This drug doesn't work like codeine, but as a local anesthetic to decrease the sensitivity of lung receptors, reducing your need to cough. The "Tess" looks like "tussive" and you can think of getting a pearl (as from an oyster) stuck in your throat. This drug is not in any way an anxiolytic (benzodiazepine), but it's interesting that the name contains the letters "b-e-n-z-o."

III. Asthma

<u>Oral steroids</u>

82. Dexamethasone (Decadron) *(DEC-ah-drawn)*

dex   = dex(terity)
a   = (comm)a
meth' = meth(od)
a   = (comm)a
sone = so(fa)(du)ne

Dexamethasone is an oral steroid with the "sone, s-o-n-e" common to many steroids.

83) **Methylpred<u>nisolone</u> (Medrol)** *(MED-rol)*

meth = meth(od)
yl   = (s)yl(lable)
pred = pred(ator)
ni'   = (k)ni(t)

A student connected "pred" and "predator of inflammation" for generic **methylprednisolone**. **Methylprednisolone** comes in a 6-day, 21-pill dose pack that gives patients 6 tablets on the first day, 5 on the 2nd, 4 on the 3rd, 3 on the 4th, 2 on the 5th, and 1 on the 6th. This dosing reminds us to *taper* steroids to allow the adrenal glands time to resume normal function.

84) **<u>Pred</u>nisone (Deltasone)** *(DEALT-uh-sewn)*

pred' = pred(ator)
ni   = (k)ni(t)
sone = so(fa)

Most students know **prednisone**, but many steroid compounds have this unofficial "sone" (s-o-n-e) ending. In this case, the "pred, p-r-e-d" is the official stem in the *prefix* position. Note, we can also find this stem in the middle of the drug name as a meaningful *infix* in the steroid methyl<u>**pred**</u>nisolone.

<u>Ophthalmic steroid</u>

85. Loteprednol ophthalmic (Lotemax) *(LOW-teh-macks)*

lo	= (Co)lo(rado)
te	= te(nt)
pred'	= pred(ator)
nol	= (etha)nol

The ophthalmic preparation **loteprednol (Lotemax)** has two o's for eyes, and comes after the two "pred" oral steroids, **methylprednisolone** and **prednisone**.

<u>Inhaled steroid / Beta$_2$ receptor agonist</u>

86. Budesonide with Formoterol (Symbicort) *(SIM-buh-court)*

bu	= bu(reau)
de'	= de(sk)
so	= so(n)
nide	= (s)nide

for	= for
mo'	= mo(at)
ter	= ter(se)
ol	= (aw)ol

Similar to the **fluticasone** paired with **salmeterol**, **budesonide** has the "sone, s-o-n-e" syllable in the middle of its name, and is pronounced "sone" even though it's spelled "s-o-n."

The "terol, t-e-r-o-l" stem in **formoterol** indicates a beta 2 agonist bronchodilator. You can think of the "S-y-m" in the brand name **Symbicort** as symbiotic, meaning "working with," plus "c-o-r-t" for corticosteroid: Sym-bi-cort.

87. Salme<u>terol</u> with Fluticasone (Advair) *(ADD-vair)*

flu = flu(id)
ti'   = ti(c)
ca   = (or)ca
sone = so(fa)(du)ne

sal = sal(ad)
me'   = me(n)
ter = ter(se)
ol   = (aw)ol

Recognize the steroid **fluticasone** by the unofficial "sone," and the long-acting bronchodilator **salmeterol** by the "terol" stem. The brand name **Advair** seems like "<u>ad</u>d two drugs to get <u>air</u>." Ad-vair.

<u>Inhaled steroid</u>

88. Budesonide (Rhinocort, Pulmicort Flexhaler)

bu   = bu(reau)
de'   = de(sk)
so   = so(n)
nide   = (s)nide

(Rhinocort, Pulmicort Flexhaler)

Budeonide bears a resemblance to **fluticasone** in that it comes as an over-the-counter nasal spray and prescription inhaler. "Rhino" is for nose and "cort" is for <u>cort</u>icosteroid in **Rhinocort**.

89. Fluticasone (Flovent HFA, Flovent Diskus, and Flonase)
(FLOW-vent)

flu   = flu(id)
ti'   = ti(c)
ca   = (or)ca
sone = so(fa)(du)ne

The "sone, s-o-n-e" ending, while not an official stem, is a useful clue that **flutica*sone*** is a steroid. In the past, inhalers felt cold when patients used them because the chlorofluorocarbon (C-F-C) propellant spray was similar to Freon, the refrigerant used in air conditioning. However, C-F-Cs damage the ozone layer, so the new ozone-safe <u>hydrofluoroalkane</u> (H-F-A) replaces the CFC propellant.

The brand name **Flovent H-F-A** cleverly uses the first two letters of the generic name **fluticasone**, incorporates f-l-o from "airflow" and adds, "vent" to let the patient know this is administered in the mouth.

The diskus, d-i-s-k-u-s inhaler is a device that looks like an Olympic discus, d-i-s-c-u-s. It employs a dry powder for inhalation rather than a propellant and liquid.

Flonase is the brand name for **fluticasone** administered nasally, and is available OTC.

Beta₂ receptor agonist short-acting

90. Albu<u>terol</u> (ProAir HFA) *(PRO-air aitch-ef-ay)*

al	= (s)al(iva)
bu'	= bu(reau)
ter	= ter(se)
ol	= (aw)ol

Albuterol's "terol, t-e-r-o-l" stem indicates it's a beta-2 adrenergic agonist that causes bronchodilation.

Note, albuterol's "terol" stem *does not* help you know if it's long acting *or* short-acting; that distinction must be memorized.

The brand name **ProAir HFA** is straightforward with "Pro" as in "I'm for it" and "Air" for airway.

Note, I discuss a newer ultra-long acting beta-2 agonist vilanterol in Chapter 8, Newer Drugs.

91. Levalbu<u>terol</u> (Xopenex HFA) *(ZOH-pen-ecks aitch-ef-ay)*

lev	= lev(el)
al	= (s)al(iva)
bu'	= bu(reau)
ter	= ter(se)
ol	= (aw)ol

Levalbuterol is the enantiomer of **albuterol**, so it should work a little bit better. I think of "<u>open</u>" and "<u>ex</u>hale" when I see the brand name **<u>X</u>open<u>ex</u>**.

Beta₂ receptor agonist / Anticholinergic

92. Albuterol with Ipratropium (DuoNeb) *(DUE-oh-neb)*

al   = (s)al(iva)
bu'   = bu(reau)
ter   = ter(se)
ol   = (aw)ol

ip   = (s)ip
ra   = (au)ra
tro'   = (me)tro
pi   = pi(ece)
um   = (g)um

A "-terol" short-acting bronchodilator like **albuterol** can combine with a shorter acting anticholinergic like *ipra*tropium (as compared to longer-acting *tio*tropium) in nebulized form to treat asthma symptoms faster.

Instructors use **atropine** as the prototype drug for the anticholinergics and you can see the "trop, t-r-o-p" stem in **atropine, ipratropium**, and **tiotropium**.

The brand name **DuoNeb** indicates a duo of drugs in nebulized form; Duo-Neb.

93. Albuterol / Ipra<u>tro</u>pium (Combivent) *(COM-bih-vent)*

al	= (s)al(iva)
bu'	= bu(reau)
ter	= ter(se)
ol	= (aw)ol
ip	= (s)ip
ra	= (au)ra
tro'	= (me)tro
pi	= pi(ece)
um	= (g)um

DuoNeb and **Combivent** have identical ingredients, but I wanted to make clear **DuoNeb** is a <u>neb</u>ulization solution and **Combivent** is a <u>comb</u>ined inhaler.

Anticholinergic

94. Tiotropium (Spiriva) *(Spur-EE-va)*

ti	= ti(e)
o	= "o"
tro'	= (me)tro
pi	= pi(ece)
um	= (g)um

Tiotropium has the same "trop, t-r-o-p" stem as **ipratropium** and is the long-acting version. As with the "terols," the beta$_2$ receptor agonists, you have to memorize which anticholinergic is long-acting versus short-acting. The brand name **Spiriva** takes the "spir, s-p-i-r" from respire, like respiration.

Just a note about the anticholinergic stems, the trop*ine* and trop*ium* indicate a difference in the number of attached atoms to the nitrogen atom. A "tropine" is a tertiary amine with three attached atoms while the "tropium" is a quaternary amine with four. I would just go ahead and memorize either the whole tropine or tropium stem respectively.

Leukotriene receptor antagonist

95. Monte<u>lukast</u> (Singulair) *(SING-you-lair)*

mon	= mon(key)
te	= te(nt)
lu'	= lu(cid)
kast	= (Out)kast

Leukotrienes that form in leukocytes (white blood cells) can cause inflammation. By blocking them, **montelukast** helps with the inflammatory component of asthma. The "-lukast, l-u-k-a-s-t" stem is similar to *leuk*otriene.

The **Singulair** brand name comes from its once daily "single" dosing and the "air" it helps to bring into the asthmatic lungs; **Singulair**.

Anti-IgE antibody

96. Omalizumab (Xolair) *(ZOHL-air)*

o	= "o"
ma	= (paja)ma
li'	= li(ly)
zu	= Zu(lu)
mab	= m(at)(c)ab

Like **infliximab** for ulcerative colitis, **omalizumab** is a biologic.

The "inf-" (i-n-f) is a prefix that separates it from other similar drugs.

The "-li-" (l-i) stands for immunomodulator (the target),

the "-zu-" (z-u) stands for humanized (the source)

and the "–mab" (m-a-b) is for monoclonal antibody.

There is a black box warning (a severe warning immediately at the beginning of the package insert) for the possibility of anaphylaxis after the first dose and even a year after the onset of treatment.

Therefore, health providers inject **omalizumab** where a medicine for treating anaphylaxis is available.

The brand name **Xolair** seems like, "exhale" and "air."

IV. Anaphylaxis

97. Epinephrine (EpiPen) *(EP-ee-pen)*

e   = (n)e(t)
pi   = pi(n)
ne'   = ne(t)
phrine = phr(ase)(femin)ine

The word **epinephrine** has a Greek origin. "Epi" means "above," and "neph" means "kidney." Above the kidney is the adrenal gland responsible for the body's natural release of **epinephrine**.

The Latinized version of **epinephrine** is adrenaline [[uh dren uh Lynn, not uh dreen uh Lynn or uh dren uh leen]] The "ad" means "above" and "renal" means "kidney"; adrenaline.

Either version stimulates the fight or flight response that can help a patient in an anaphylactic state.

The **EpiPen** brand name comes from the injector device that looks somewhat like a pen.

CHAPTER 4 - IMMUNE

I. OTC Antimicrobials

<u>Antibiotic cream</u>

98. Neomycin /Polymyxin B /Bacitracin, Neosporin.
(KNEE-oh-spore-in)

ba   = ba(t)
ci   = (s)ci(ntillate)
tra'   = tra(y)
cin   = cin(der)

ne   = (mo)ne(y)
o   = "o"
my'   = my
cin   = cin(der)
sul'   = sul(len)
fate   = fate

po   = po(t)
ly   = (li)ly
myx'   = myx(edema)
in   = in
B   = "B"

The **neomycin** component is an aminoglycoside generally toxic to the kidney (nephrotoxic) and ears (ototoxic) when used systemically (in the body). However, patients safely use *topical* preparations containing **neomycin** such as over-the-counter **Neosporin**. The brand **Neosporin** takes "N-e-o and 's'" from **neomycin sulfate**, "p-o" from **polymyxin B**, and "r-i-n" from **bacitracin**."

99. Mupirocin (Bactroban) [RX] *(BACK-troh-ban)*

mu   = mu(sic)
pi'   = pi(ece)
ro   = ro(w)
cin   = cin(der)

I added the prescription antibiotic, **mupirocin (Bactroban)** for impetigo after OTC Neosporin because it is prescription-only.

Antifungal cream

100. Butenafine (Lotrimin Ultra) *(LOW-treh-min)*

bu   = bu(reau)
te'   = te(nt)
na   = (bana)na
fine   = fi(eld)(li)ne

Butenafine treats topical fungal infections like ringworm, jock itch, and athlete's foot. Sometimes you will see the Latin names for these conditions: *tinea coporis* (ringworm), *tinea cruris* (jock itch), and *tinea pedis* (athlete's foot) respectively.

101. Terbinafine (Lamisil) *(LAM-ih-sill)*

ter	= ter(se)
bi'	= bi(t)
na	= (bana)na
fine	= fi(eld)(li)ne

I alphabetically listed the antifungal **terbinafine** after **butenafine (Lotrimin Ultra)**. I didn't see a formal stem, but both end with n-a-f-i-n-e.

102. Clotrimazole / Betamethasone (Lotrisone) [RX] *(LOW-trih-sewn*

clo	= clo(ver)
tri'	= tri(p)
ma	= (paja)ma
zole	= (fe)z(p)ole
be	= be(ta)
ta	= (da)ta
meth'	= meth(od)
a	= (comm)a
sone	= so(fa)(du)ne

By adding **betamethasone**, a steroid, to **clotrimazole,** an OTC antifungal, the combination becomes prescription-only **Lotrisone**.

Vaccinations [Some RX, some antibacterial]

I listed the vaccinations around the influenza vaccine. Some require a prescription (depending on varying state laws) and some are antibacterial. I alphabetized them by generic name. Since most vaccinations have the agent they prevent against in their names, mnemonics don't really make sense.

103. Diphtheria toxoid (Boostrix)

104. Haemophilus influenzae Type B (Pedvax HIB)

105. Influenza Vaccine (brand names Fluzone, and Flumist)
(FLEW-zone, FLEW-mist)

in = in
flu = flu(id)
en' = (p)en
za = (pla)za

vac' = vac(uum)
cine = (leu)cine

Some children may need a prescription for **influenza vaccine**, most adults can readily get flu shots. Careful; generic names with "f-l-u." **Fluconazole,** an antifungal, and **fluoxetine**, an antidepressant, refer to fluorine atoms in their chemical structure, not influenza virus.

The "flu" in the vaccine brand names **Fluzone** and **Flumist** indicates in<u>flu</u>enza, just like the brand **Tamiflu**, an oral medication for influenza infection that contains the syllable "flu." The *nasal* vaccine **Flumist** provides an alternative for patients who don't want an injection. However, the CDC Advisory Committee on Immunization Practices (ACIP) voted to not use the "nasal spray" flu vaccine during the 2016-2017 season because of data showing poor effectiveness from 2013-2016. Either way, knowing if a drug is brand or generic becomes critical to knowing the meaning of f-l-u.

106. Measles, Mumps and Rubella (MMR)

107. Meningococcal (conjugate and polysaccharide) Menomune

108. Pertussis in combination

109. Pneumococcal (conjugate and polysaccharide) (Prevnar 13, Pneumovax 23)

110. Polio

111. Rotavirus (RotaTeq)

112. Tetanus in combination

113. Varicella (Varivax)

114. Zoster (Zostavax)

Note, **Varivax** and **Zostavax** prevent varicella (chickenpox) and herpes zoster (shingles) respectively. The "vax" in the brand names indicates "vaccine."

Antiviral OTC

115. Docosanol (Abreva) *(uh-BREE-vah)*

do   = do(ugh)
co'   = co(ne)
sa   = (vi)sa
nol   = (etha)nol

Docosanol is a topical antiviral for cold sores caught and treated early. I thought, "who would pay twenty dollars for a small tube like that?" Then I thought of homecoming dances and ruined pictures with a large cold sore.

So, my mnemonic became "use **docosanol**, so you can go to the ball."

Abreva, the brand name, hints at therapeutic effect as **Abreva** "<u>abbrev</u>iates" the time a cold sore lasts.

II. Antibiotics that affect the cell wall

<u>Penicillins</u>

116. Amoxi<u>cillin</u> (Amoxil) *(uh-MOCKS-ill)*

a   = (comm)a
mox   = mox(ie)
I   = i(t)
cil'   = (pen)cil
lin   = lin(t)

Amoxicillin has the "cillin, c-i-l-l-i-n" stem that indicates its relationship to the penicillin family.

The "a-m-o" probably came from the fact that it's an "<u>amino</u>" penicillin.

The "cillin" stem sounds like "*cell*-in" and can help you remember that **amoxicillin** or, more generally, **penicillins** destroy the *cell* wall.

The brand name **Amoxil** simply removes an "i-c" and "l-i-n" from the generic name, amoxicillin, to make Amoxil.

117. Peni<u>cillin</u> (Veetids) *(Veetids)*

pe   = pe(n)
ni   = (k)ni(t)
cil'   = (pen)cil
lin   = lin(t)

Follows **amoxicillin** alphabetically with the same "cillin" stem.

Penicillin/Beta-lactamase inhibitor

118. Amoxicillin with clavulanate (Augmentin) *(awg-MENT-in)*

a	= (comm)a
mox	= mox(ie)
i	= i(t)
cil'	= (pen)cil
lin	= lin(t)
cla'	= cla(w)
vu	= (re)vu(e)
la	= la(wn)
nate	= (in)nate

When **amoxicillin** alone doesn't work, we can think of the brand **Augmentin** as it augments **amoxicillin's** defenses against the bacterial beta-lactamase enzyme with **clavulanate**.

I think of the "clavicle," the bone in your shoulder, as protective of the upper lung and associate **clavulanate** with that same protective effect as both start with c-l-a-v.

Cephalosporins [by generation]

119. <u>Cephalexin</u> (Keflex) 1st generation *(KEH-flecks, not KEY-flecks)*

ce	= ce(nt)
pha	= (al)pha
lex'	= (f)lex
in	= in

With cephalosporins, a newer generation has better properties than the last, relative to what the prescriber is treating. Those thee advantages include 1) better penetration into the cerebrospinal fluid (C-S-F), 2) better gram- negative coverage, and 3) better resistance to beta-lactamases.

You may lose some gram-positive coverage as you move up the spectrum, however. The "ceph, c-e-p-h" is an old stem from the first generation. The new stem "cef, c-e-f) identifies the newer generations.

That's how I remember **cephalexin** as first generation. The brand name **Keflex, which begins with a "K,"** takes some letters from **cephalexin** to make its name.

120. <u>Cef</u>uroxime (Ceftin) *(SEF-tin)*

cef	= c(l)ef
u'	= "u"
rox	= (xe)rox
ime	= (ox)ime

Is 2nd generation and has the cephalosporin prefix c-e-f.

121. Cefdinir (Omnicef) *(AHM-knee-sef)*

```
cef'   = c(l)ef
di     = di(nner)
nir    = (souve)nir
```
(AHM-knee-sef)

is 3rd generation and has the cephalosporin prefix c-e-f.

122. Ceftriaxone (Rocephin) *(row-SEF-in)*

```
cef    = c(l)ef
tri    = tri(angle)
ax'    = ax
one    = (c)one
```

In the generic name, **ceftriaxone's** "cef-" (c-e-f) indicates it's a cephalosporin. There is a "tri" (t-r-i) in the generic name that you can use to remember it is 3rd generation.

Rocephin, the brand name, seems to come from Hoffman-LaRoche's patent.
The drug company took the "Ro" (r-o) from "LaRoche," and the "ceph" (c-e-p-h) and "in" from cephalosporin to make **Ro-ceph-in**.

123. Cefepime (Maxipime) *(MAX-eh-peem)*

```
cef'   = c(l)ef
e      = (en)e(my)
pime   = pi(ece)(li)me
```

Cefepime is a fourth-generation cephalosporin. I've remembered it by thinking of four letters "p-i-m-e" that are in both the brand name **Maxipime** and the generic name **cefepime**. Also, at the time, **Maxipime** was the *maximum* generation, the fourth and highest. However now there is a fifth-generation cephalosporin, ceftaroline, but that old mnemonic stayed in my brain.

Glycopeptide

124. Vancomycin (Vancocin) *(VAN-co-sin)*

van   = van
co   = co(ne)
my'   = my
cin   = cin(der)

Vancomycin's "mycin, m-y-c-i-n" stem isn't very useful for finding its therapeutic class.

All "mycin" really means is that chemists derived **vancomycin** from the *Streptomyces* bacteria, taking the m-y-c from *Streptomyces*. I remember the function as "**vancomycin** will vanquish MRSA." To remember the brand name **Vancocin**, remove the "my" from **vancomycin,** to get V-a-n-c-o-c-i-n. **Vancocin.**

III. Antibiotics - Protein Synthesis Inhibitors (Bacteriostatic)

Tetracyclines

125. Doxycycline (Doryx) *(DOOR-icks)*

dox = (para)dox
y = (countr)y
cy' = cy(cle)
cline = cli(ff)(vi)ne

I use the "d" in **doxycycline** to remind me that dentists use it to treat periodontal disease.

Doryx, the brand name, takes the first four letters of **doxycycline** and adds an "r," scrambling them a bit to create Doryx.

126. **Minocycline (Minocin)** *(MIN-oh-sin)*

mi = mi(nt)
no = (can)no(n)
cy' = cy(cle)
cline = cli(ff)(vi)ne

Minocycline, like **doxycycline** has the **tetracycline** class "cycline, c-y-c-l-i-n-e" stem. To create the brand name **Minocin**, the manufacturer dropped the "c-y-c-l" and last "e," from minocycline to get M-i-n-o-c-i-n, Minocin.

127. **Tetracycline (Sumycin)** *(sue-MY-sin)*

te = te(nt)
tra = (orches)tra
cy' = cy(cle)

Tetracycline, although it's essentially unavailable, would follow alphabetically after **doxycycline** and **minocycline**, but it's the class' namesake so I included it.

Macrolides

128. Azi<u>thromy</u>cin (brand names Zithromax or Z-Pak) *(ZITH-row-max)*

a   = (comm)a
zi   = zi(pper)
thro   = thro(w)
my'   = my
cin   = cin(der)

To recognize the three macrolides, **azithromycin, clarithromycin**, and **erythromycin**, you will see the "mycin, m-y-c-i-n" ending, but also a possible infix of "thro, t-h-r-o."

Be careful: there *are* macrolides without this infix and stem. The brand name **Zithromax** takes seven letters from **azithromycin** to construct its name. The **Zithromax Z-pak** is a convenient six-tablet package that includes a five-day course, two tablets for a loading dose on day one and one tablet for each thereafter.

129. Clari<u>thromy</u>cin (Biaxin) *(bi-AX-in*

cla   = cla(rinet)
ri   = ri(nse)
thro   = thro(w)
my'   = my
cin   = cin(der)

Again, you can use the "thromycin" to help clue you in to this drug's class. Gastroenterologists prescribe **clarithromycin** for peptic ulcer disease (PUD) triple therapy along with **amoxicillin** and a proton pump inhibitor like **omeprazole**. The "b-i" in the brand name **Biaxin** indicates the twice daily dosing from the Latin abbreviation b.i.d. for *bis in die*; twice a day.

130. Ery<u>thromy</u>cin (E-Mycin) *(E-MY-sin)*

e	= (en)e(my)
ry	= (o)ry(x)
thro	= thro(w)
my'	= my
cin	= cin(der)

Some **erythromycin** tablets are bright red and that might be where it got its name. An erythrocyte is a red blood cell, and the word comes from connecting "erythro," the Greek for "red," and "cyte," for cell.

The brand name **E-mycin** comes from taking the "rythro" out of the generic name **erythromycin** to render **E-mycin**.

131. Fida<u>xo</u>micin (Dificid) *(DIF-ih-sid)*

fi	= fi(n)
dax'	= (a)dax(ial)
o	= "o"
mi'	= mi(le)
cin	= cin(der)

The "dax" in the name **fi<u>dax</u>omicin (Dificid)** might come from its source, *Dactylosporangium aurantiacum*. The brand name **Dificid** indicates its primary therapeutic use against *Clostridium di<u>ffici</u>le*.

Lincosamide

132. Clindamycin (Cleocin) *(KLEE-oh-sin)*

clin	= clin(ic)
da	= (so)da
my'	= my
cin	= cin(der)

Most students remember the adverse effect C-DAD (Clostridium difficile-Associated Disease) because there is a "c" and a "da" right after each other in the generic name **clindamycin**.

To make the brand name **Cleocin**, the manufacturer replaced the "i-n-d-a-m-y" in **clindamycin** with "e-o," for Cleocin.

Oxazolidinone

133. Line<u>zolid</u> (Zyvox) *(ZIE-vocks)*

li   = li(ly)
ne'   = ne(t)
zo   = (hori)zo(n)
lid   = lid

The "zolid, z-o-l-i-d" stem in **linezolid** comes from the **oxa<u>zolid</u>inone** class.

I think it's more helpful to think, "Man, **Zyvox** is zolid (solid); it treats two very difficult-to-kill organisms, Methicillin resistant staph aureus, M-R-S-A and vancomycin resistant enterococci, V-R-E."

IV. Antibiotics - Protein Synth. Inhibitors (Bactericidal)

<u>Aminoglycosides</u>

I think of the "side" (s-i-d-e) in **aminoglycoside** and "cide, c-i-d-e" as in "cidal" to remind me these drugs are killers.

134. Ami<u>ka</u>cin (Amikin) *(AM-eh-kin)*

a	= (m)a(t)
mi	= mi(nt)
ka'	= (s)ka(te)
cin	= cin(der)

Some internet sources say that a "cin, c-i-n" ending means an aminoglycoside, but that is not necessarily true. Many antibiotics end in "c-i-n," so I think it's more useful to look at the "a-m-i" that is in **aminoglycoside** *(the drug class)*, **ami**kacin *(the generic name)* and **Amikin (the brand name)**.

The brand name **Amikin** is simply **amikacin** without the "a-c." **Amikin.**

135. Genta<u>mi</u>cin (Garamycin) *(gare-uh-MY-sin*

gen	= gen(tleman)
ta	= (da)ta
mi'	= mi(le)
cin	= cin(der)

Just as practitioners abbreviate **vancomycin** as "vanc" in conversation, they abbreviate **gentamicin** as "gent." The brand name **Garamycin** is similar to **gentamicin** spelled with "mycin, m-y-c-i-n" not "micin, m-i-c-i-n". This World Health Organization frowns on brand names that have confusing stem-like parts to them.

V. Antibiotics for Urinary Tract Infections (UTIs) and Peptic Ulcer Disease (PUD)

<u>OTC Urinary tract analgesic</u>

136. Phenazopyridine (Uristat) *(YOUR-ih-stat)*

phe = phe(nomenon)
na' = na(p)
zo   = zo(ne)
py' = py(lon)
ri   = ri(nse)
dine = (bir)di(e)(tu)ne

Phenazopyridine is an over-the-counter urinary tract analgesic that allows a patient to get some relief before she can see her physician for treatment of a bladder infection.

Nitrofuran

137. <u>Ni</u>tro<u>fur</u>antoin (Macrobid, Macrodantin) *(MACK-row-bid, mack-row-DAN-tin)*

ni	= ni(ght)
tro	= (me)tro
fur'	= fur(y)
an	= an
to	= to(e)
in	= in

Nitrofurantoin is a nitrofuran antibiotic, as the first brand name implies, and is taken twice daily as indicated by the "b-i-d" in the name.

Dihydrofolate reductase inhibitor

138. Sulfamethoxazole with Trimethoprim (Bactrim) *(BACK-trim)*

sul   = sul(len)
fa   = (so)fa
meth   = meth(od)
ox'   = ox
a   = (comm)a
zole   = (fe)z(p)ole

tri   = tri(angle)
meth'   = meth(od)
o   = "o"
prim   = prim

S-M-Z forward slash **T-M-P** is the acronym for **sulfamethoxazole** with **trimethoprim**. The "sulfa, s-u-l-f-a" in sulfamethoxazole and the "prim, p-r-i-m," in trimethoprim put them in the dihydrofolate reductase inhibitor class. (Remember this enzyme helps the bacteria make folic acid) While sulfa drugs have "s-u-l-f-a" in them, some drugs have sulfa groups in the chemicals' structure, but not in the generic name, e.g. [[for example]], **furosemide.** I want to caution you about seeing sulfa moieties in chemical structures and assuming it will cause an allergic reaction. The academic literature doesn't support cross-sensitivity between allergies to sulfa antibiotics and other sulfonamide containing drugs like **furosemide**. The brand name contains "b-a-c-t-r-i-m" from "**bacterium**" to make **Bactrim**.

Fluoroquinolones

139. Ciprofloxacin (Cipro) *(SIP-row)*

ci   = (s)ci(ntillate)
pro   = pro(ton)
flox' = fl(ocks)ox(en)
a   = (comm)a
cin   = cin(der)

The **quinolone** stem is "oxacin, o-x-a-c-i-n" but **fluoroquinolones** have the "f-l" infix also.

One student remembered quinolones are for UTIs because Dr. *Quinn*, Medicine Woman, a 90's television doctor, is female. Women get proportionally more UTIs than men do. By cutting the "floxacin" stem from the generic name **ciprofloxacin**, the manufacturer made the brand name, **Cipro**.

140. Gatifloxacin ophthalmic (Zymar) *(ZIE-mar)*

ga   = ga(s)
ti   = ti(c)
flox' = fl(ocks)ox(en)
a   = (comm)a
cin   = cin(der)

I alphabetically added **gatifloxacin ophthalmic** after ciprofloxacin, with the same -floxacin stem.

141. Levofloxacin (Levaquin) *(LEV-uh-quinn)*

le   = le(tter)
vo   = vo(te)
flox' = fl(ocks)ox(en)
a   = (comm)a
cin = cin(der)

Levofloxacin is the left-handed (levo) isomer of **ofloxacin,** another **fluoroquinolone**. The brand name combines the "lev" from **lev**ofloxacin and "quin" from fluoro**quin**olone to form **Levaquin**.

142. Moxifloxacin (Avelox, Vigamox) *(AV-ah-locks, VIH-gah-mocks)*

mox = mox(ie)
i   = i(t)
flox' = fl(ocks)ox(en)
a   = (comm)a
cin = cin(der)

Vigamox is an ophthalmic preparation that contains "v-i" for vision and moxifloxacin has the same –floxacin stem.

Nitroimidazole

143) Metro<u>nidazole</u> (Flagyl) *(FLADGE-ill)*

me	= me(n)
tro	= (me)tro
ni'	= (k)ni(t)
da	= (so)da
zole	= (fe)z(p)ole

The generic name **metronidazole** differs by only three letters from the word "nitroimidazole," its parent class. It's stem, "nidazole, n-i-d-a-z-o-l-e," is sometimes confused as "a-z-o-l-e" which is simply a type of chemical compound. Make sure to use the whole stem, otherwise you'll confuse the other stems that end in "a-z-o-l-e" like prazole, conazole, and piprazole.

Gastroenterologists use **metronidazole** for peptic ulcer disease (P-U-D). Note that **metronidazole** is technically an antiprotozoal. Some students look at the "azole" (a-z-o-l-e) ending not as a stem, but for its similarity to "ozoal" (o-z-o-a-l) from "protozoal." Another student learned to give "Flag,"[[pronounce fladge]] a shorter form of metronidazole's brand **Flagyl**, for *B. frag,* [[pronounce Bee Frag]] a shortening of the *Bacteroides fragilis* infections.

VI. Anti-tuberculosis agents

144) R<u>if</u>ampin (Rifadin) *(RIFF-ah-din)*

ri = ri(nse)
fam' = fam(ily)
pin = pin

Students remember that **rifampin** turns secretions like tears, sweat, and urine red with its first letter "r." The brand name **Rifa<u>d</u>in** simply replaces the "m-p" from **rifampin** with a "d" to get Rifadin.

145) Isoniazid (INH) *(EYE-en-aitch)*

i = "i"
so = so(n)
ni' = ni(ght)
a = (comm)a
zid = zi(ppe)d

The "n-i" in the middle of **iso<u>ni</u>azid** reminds students that peripheral <u>n</u>eur<u>i</u>tis is an adverse effect, taking the "n-i" from neuritis. There is no "H" in **isoniazid**, so the brand name **I-N-H** comes from those letters in the chemical name <u>i</u>so<u>n</u>icotinyl<u>h</u>ydrazide.

146) Pyrazinamide (P-Z-A) *(pee-zee-ay)*

py = py(lon)
ra = (au)ra
zi' = zi(pper)
na = (bana)na
mide = (gu)m(sl)ide

The "p" in **pyrazinamide** reminds students that an adverse effect is polyarthritis. Pulling letters from within the generic name, **py<u>ra</u>z<u>i</u>n<u>a</u>mide's** abbreviation is **P-Z-A**.

147) Ethambutol (Myambutol) *(my-AM-byou-tall)*

eth   = (m)eth(od)
am'   = (l)am(p)
bu   = bu(reau)
tol   = tol(erate)

The "e" for "eyes" or "o" in **ethambutol** helps remind students of vision and optics, and that optic neuritis is an adverse effect.

To make the brand name, the manufacturer replaced the "eth" in **ethambutol** with "my." The m-y begins **Myambutol** because *Mycobacterium tuberculosis* is the causative agent.

VII. Antifungals

148. Amphotericin B (Fungizone) *(FUN-gah-zone)*

am   = (l)am(p)
pho   = pho(ne)
te'   = te(nt)
ri   = ri(nse)
cin   = cin(der)

What about **amphotericin A**? Well, it didn't do anything, so they came up with **amphotericin B**. While the antibacterials' brand names didn't do a very good job helping to indicate their therapeutic effects, this antifungal's brand name, **Fungizone**, makes it easier to know its therapeutic use.

149. Fluconazole (Diflucan) *(die-FLUE-can)*

flu   = flu(id)
co'   = co(t)
na   = (bana)na
zole = (fe)z(p)ole

The "conazole, c-o-n-a-z-o-l-e" ending helps identify **fluconazole** as an antifungal drug. Again, be careful of the "f-l-u" in **fluconazole**, which is for a fluorine atom it contains, not influenza.

One student came up with using the first three letters of the brand name **Diflucan** as the vehement command, "**Di**e fungi!"

150. Ketoconazole (Nizoral) *(nye-ZOR-ahl)*

ke   = ke(y)
to   = (pis)to(n)
co'   = co(t)
na   = (bana)na
zole = (fe)z(p)ole

The antifungal **ketoconazole** fits alphabetically with the other "azole"[[uh-zole]] antifungal **fluconazole**.

151. Nystatin (Mycostatin) *(MY-co-stat-in)*

ny'   = ny(lon)
sta   = sta(ff)
tin   = tin

Nystatin is an interesting generic name because it ends in "statin." A class of cholesterol lowering drugs, the HMG-CoA reductase inhibitors, commonly referred to as "statins," have a similar ending.

A better substem + suffix stem for HMG-CoA reductase inhibitors is "vastatin, v-a-s-t-a-t-i-n."

To keep from thinking **nystatin** was ever a cholesterol lowering "statin," one student remembered the dosage forms nystatin comes in: powder and liquid to swish and spit.

Mycostatin, the brand name, dropped the "ny, n-y" from nystatin and added "Myco, m-y-c-o" a prefix often seen with mycoses (fungal infections).

VIII. Antivirals - Non-HIV

Influenza A and B

152. Oseltamivir (Tamiflu) *(TA-mi-flue)*

o   = "o"
sel   = sel(fie)
ta'   = ta(b)
mi   = mi(nt)
vir   = vir(ulent) [rhymes with veer]

Recognize oseltamivir's drug class from the amivir, a-m-i-v-i-r stem. Often family members will all get prescriptions for **oseltamivir** if one person is sick enough or if a family member is immunocompromised. The brand name **Tamiflu** alludes to a drug that "tames the flu." It's prescribed for acute influenza or prophylaxis.

153. Zanamivir (Relenza) *(rah-LEN-zuh)*

za   = (pla)za
na' = na(p)
mi   = mi(nt)
vir = vir(ulent) [rhymes with veer]

Also recognize zanamivir's drug class from the amivir, a-m-i-v-i-r stem. **Zanamivir** comes in a Diskhaler, a way to get powder to the lungs. The Diskhaler is difficult for patients with dexterity issues, but provides an alternative to **oseltamivir (Tamiflu).**

Think: **Relenza** "represses influenza" virus, or **Relenza** makes "influenza relent" (give up).

Herpes simplex virus & Varicella-Zoster Virus HSV/VSV

154. Acyclovir (Zovirax) *(ZOH-vih-racks)*

a	= "a"
cy'	= cy(cle)
clo	= clo(ver)
vir	= vir(ulent) [rhymes with veer]

Zovirax treats Varicella-Zoster virus (VZV) and herpes simplex virus (HSV). You can think of **Zovirax** as a drug that axes Zoster virus. Dosing is five times daily.

155. Valacyclovir (Valtrex) *(VAL-trex)*

val	= val(ley)
a	= (comm)a
cy'	= cy(cle)
clo	= clo(ver)
vir	= vir(ulent) [rhymes with veer]

Valacyclovir has **acyclovir** in the name because it's the valine ester. A prodrug like **valacyclovir** turns into an active drug in the body, in this case, **acyclovir**. **Valacyclovir** allows for twice daily dosing, so prescribers prefer the oral form to five times daily dosing of **acyclovir** for patient compliance. The brand name, **Valtrex,** includes the "val" from **valacyclovir** plus T-rex, and wrecks a virus; Val-T-Rex... Valtrex.

Respiratory Syncytial Virus RSV

156. Palivizumab (Synagis) *(SIN-uh-giss)*

pa	= pa(th)
li	= li(ly)
vi'	= vi(sion)
zu	= Zu(lu)
mab	= m(at)(c)ab

The prefix "p-a-l-i" has "p" and "i" in it. You can remember **palivizumab** is for pediatrics or infants at risk for RSV. In **palivizumab**, the "pali" is a prefix that separates it from similar drugs.

The "-vi-" (v-i) stands for antiviral (the target), the "-zu-" (z-u) stands for humanized (the source), and the "–mab" (m-a-b) is for monoclonal antibody.

This biologic stem + infixes resemble **infliximab (Remicade)** for ulcerative colitis or **omalizumab (Xolair)** for asthma, but with a different clinical purpose.

Hepatitis

157. Ente<u>cavir</u> (Baraclude) *(BARE-ah-clude)*

en	= (p)en
te'	= te(nt)
ca	= (or)ca
vir	= vir(ulent) [rhymes with veer]

Entecavir has the "v-i-r, vir" common to antivirals and c-a-v-i-r, cavir stem. Entecavir is an anti-Hepatitis B virus reverse transcriptase inhibitor

158. Hepatitis A (Havrix) *(HAVE-ricks)*

159. Hepatitis B (Recombivax HB) *(re-CAHM-bah-vax)*

Entecavir is for active hepatitis infections, while both **Hepatitis A (Havrix)** and **Hepatitis B (Recombivax HB)** vaccines are preventative. Learn more about the Hepatitis C drugs in Chapter 8.

HPV

160. Human papillomavirus (Gardasil) *(GAR-dah-sil)*
Gardasil gu<u>ard</u>s against human papillomavirus (HPV).

IX. Antivirals - HIV

<u>Fusion Inhibitor</u>

161. Enfu<u>vir</u>tide (Fuzeon) (ENF, T-20) *(FYOO-zee-on)*

en	= (p)en
fu'	= fu(mes)
vir	= vir(tue)
tide	= tide

It's easier to remember **enfuvirtide's** brand name **Fuzeon** first because it's a <u>fusion</u> inhibitor, taking the f-u and o-n from fusion; **Fuzeon**. Inside the generic name, you see "vir" (v-i-r) for antiviral and we pronounce the "f-u" as FYOO. Put that together and you can remember **enfuvirtide** is brand name **Fuzeon**, a fusion inhibitor. I use the "T" in "**T-20**" to remember that "tide" is the last syllable in the generic name.

CCR5 Antagonist

162. Mara<u>vir</u>oc (Selzentry) (MVC) *(SELL-zen-tree)*

ma	= (paja)ma
ra	= ra(ttle)
vi'	= vi(sion)
roc	= roc(k)

The brand name **Selzentry** sounds a lot like "sentry," someone who guards. The stem "vir, v-i-r" is inside the generic name mara<u>vir</u>oc. The sub-stem "viroc, v-i-r-o-c" has five letters, with the "c" at the end of the generic name, so you can remember the two "c's" in CCR5 antagonist. You can think of **maraviroc** as a "<u>roc</u>k" guarding against viral entry.

Non-nucleoside reverse transcriptase inhibitors (NNRTI) with two nucleoside / nucleotide reverse <u>transcriptase</u> inhibitors (<u>NRTIs</u>)

163. Efa<u>vir</u>enz (Sustiva) [NNRTI] *(suss-TEE-vah)*

e	= (n)e(t)
fa'	= fa(ct)
vir	= vir(tue)
enz	= enz(yme)

164. Emtri<u>ci</u>tabine / Tenofo<u>vir</u> (Truvada) [NRTIs] *(true-VAH-dah)*

em	= (g)em
tri	= tri(angle)
ci'	= ci(der)
ta	= (da)ta
bine	= (zom)bi(e)(li)ne
te	= te(nt)
no'	= no(d)
fo	= (ef)fo(rt)
vir	= vir(ulent) [rhymes with veer]

I put the HIV medications in category order. **Efavirenz** is an NNRTI, and both **emtricitabine** and **tenofovir** are nucleotide reverse transcriptase inhibitors (NRTIs). Those three medications are the same medications that are in **Atripla**.

165) Efavirenz with Emtricitabine and Tenofovir (Atripla) (EFV with FTC and TDF) *(ay-TRIP-lah)*

e	= (n)e(t)
fa'	= fa(ct)
vir	= vir(tue)
enz	= enz(yme)
em	= (g)em
tri	= tri(angle)
ci'	= ci(der)
ta	= (da)ta
bine	= (zom)bi(e)(li)ne
te	= te(nt)
no'	= no(d)
fo	= (ef)fo(rt)
vir	= vir(ulent) [rhymes with veer]

The brand name **Atripla** can be thought of as three drugs, "triple" surrounded by two A's that can stand for "against AIDS." When you see something complex to memorize, first look at the sub-class of antiviral. Instead of trying to memorize the whole name, try to memorize the stems "virenz, v-i-r-e-n-z," "citabine, c-i-t-a-b-i-n-e," and "vir, v-i-r". Then add the other two or three syllables to memorize the whole names of **efavirenz, emtricitabine,** and **tenofovir.**

Integrase Strand Transfer Inhibitor

166. Ral<u>tegra</u>vir (Isentress) (RAL) *(EYE-sen-tress)*

ral   = ral(ly)
te'   = te(nt)
gra   = (ag)gra(vate)
vir   = vir(ulent) [rhymes with veer]

The brand **Isentress** also looks like sentry, except it has the "I" to reference the integrase enzyme. **Raltegravir** is an integrase strand transfer inhibitor. Inside the generic name, you find the stem "tegravir, t-e-g-r-a-v-i-r" made of "tegra, t-e-g-r-a" a part of in<u>tegra</u>se and "vir, v-i-r" for anti<u>vir</u>al.

Protease Inhibitor

167. Atazanavir (Reyataz) (ATV) *(RAY-ah-taz)*

a   = (m)a(t)
ta   = (da)ta
za' = za(p)
na   = (bana)na
vir   = vir(ulent) [rhymes with veer]

The protease inhibitor **atazanavir** fits in alphabetically before darunavir, the other protease inhibitor.

168. Darunavir (Prezista) (DRV) *(Pre-ZIST-uh)*

da   = (so)da
ru' = ru(by)
na   = (bana)na
vir   = vir(ulent) [rhymes with veer]

The brand **Prezista** sounds like resist if it was spelled "r-e-z-i-s-t," and the first two letters "p-r" of "protease." The "navir, n-a-v-i-r" stem indicates protease inhibitor.

X. Miscellaneous

169. Albendazole (Albenza) *(all-BENZ-ah)*

al   = (s)al(iva)
ben' = ben(d)
da   = (so)da
zole = (fe)z(p)ole

is an anthelmintic, which means "against worms." Careful, those other YouTube videos, not mine, that said –azole was a proton pump inhibitor or antifungal, can get you into trouble here. The stem is –bendazole, b-e-n-d-a-z-o-l-e.

170. Hydroxychloroquine (Plaquenil) *(PLAK-quenn-ill)*

hy   = hy(brid)
drox = dr(ink)(b)ox
y   = (countr)y
chlor' = chlor(ine)
o   = "o"
quine = quin(c)e

Is an antimalarial.

171. Nitazoxanide (Alinia) *(ah-LIH-knee-ah)*

ni'   = ni(ght)
ta   = (da)ta
zox' = (qui)z(b)ox
a   = (comm)a
nide = (s)nide

is an antiprotozoal. The first three letters in n-i-t-a can be unscrambled to make a-n-t-i, anti.

Chapter 5 - Neuro

I. OTC Local anesthetics and antivertigo

<u>Local anesthetics</u>

172. Benzo<u>caine</u> (Anbesol) (ANN-buh-sawl)

ben'   = ben(d)
zo   = zo(ne)
caine = c(at)(migr)aine

The "caine, c-a-i-n-e" stem indicates **benzocaine** is a local anesthetic.

Anbesol has the letter "n" and the letter "b" in it, the first and last letter of the word <u>n</u>um<u>bs</u> for its use in numbing an aching tooth.

173. Lido<u>caine</u> (OTC brand, Solarcaine) *(SOH-ler-cane)*

li'   = li(ght)
do   = do(ugh)
caine = c(at)(migr)aine

We often use **lidocaine** topically over-the-counter to treat sunburns, hence the brand name **Solarcaine.**

Injectable and patch forms of **lidocaine** are also available by prescription. Paramedics often use **lidocaine** for arrhythmias in emergencies as an injectable, which is part of the Lean, L-E-A-N acronym for the emergency medicines: **<u>l</u>idocaine**, **<u>e</u>pinephrine**, **<u>a</u>tropine**, and **<u>n</u>aloxone**.

Antivertigo

174. Meclizine (Dramamine) *(DRAH-mah-mean)*

me'   = me(n)
cli   = cli(ff)
zine   = (maga)zine

If you squish the "c" and "l" in meclizine to make a "d" and use the ensuing "i-z-i," you get d-i-z-i, dizzy. **Antivert** is an alternate brand name to Dramamine and forms part of the word **antivert**igo. There's one more over-the-counter combination medication before we segue into prescription-only products. Acetaminophen with diphenhydramine is better known stateside as brand Tylenol P-M.

II. Sedative-hypnotics (Sleeping pills)

<u>OTC Non-narcotic analgesic / Sedative-hypnotic</u>

175. Acetaminophen/ Diphenhydramine (Tylenol PM) *(TIE-len-all pee em)*

a	= (comm)a
ce	= ce(iling)
ta	= (da)ta
mi'	= mi(nt)
no	= (can)no(n)
phen	= (hy)phen

di	= di(ce)
phen	= (hy)phen
hy'	= hy(brid)
dra	= (hy)dra
mine	= (hista)mine

It's common to combine two drugs like **acetaminophen** for aches and **diphenhydramine**, a sedating 1st- generation antihistamine. One of my students came up with mnemonic to take both "phens," acetamino*phen* and di*phen*hydramine when you want to end up sleepin'.

<u>Benzodiazepine-like</u>

176. Eszopi<u>clone</u> (Lunesta) *(Lou-NES-tuh)*

es = (m)es(s)
zo = zo(ne)
pi' = pi(n)
clone = clone

Eszopiclone's generic stem "-clone, c-l-o-n-e" will put you in the sleeping zone is how one student remembered eszopiclone's function.

Some students point to the "z" in **es<u>z</u>opiclone** for getting your z's.

The brand name **Lunes<u>ta</u>** uses the Latin for moon (Luna) plus part of the word "rest," which combines a visual with therapeutic function. **Lunesta**.

177. Zol<u>pidem</u> (Ambien or Ambien CR) *(AM-bee-en)*

zol' = zo(na)l
pi = pi(n)
dem = dem(ocracy)

Use the "-pidem, p-i-d-e-m" stem to remember **zol<u>pidem</u>** as a sedative-hypnotic. Some students point to the "z" in **zolpidem** for getting z's just like eszopiclone. Some students match the brand name **Ambien** with an <u>ambien</u>t, sleepy environment. **Zolpidem** <u>C</u>ontrolled <u>R</u>elease , brand **Ambien C-R**, works for people who have difficulty maintaining sleep (D-M-S) *and* those who have difficulty falling asleep (D-F-A).

The regular zolpidem only works for those with difficulty falling asleep, D-F-A.

<u>Melatonin receptor agonist</u>

178. Ra<u>melteon</u> (Rozerem) *(row-ZER-em)*

ra	= (au)ra
mel'	= mel(on)
te	= te(a)
on	= on

The "–melteon, m-e-l-t-e-o-n" stem in **ra<u>melteon</u>** lets you know it's a melatonin agonist.

A student said the "m-e-l" in **ra<u>mel</u>teon** reminded her of the word <u>mel</u>low, m-e-l-l-o-w

In **Rozerem**, you can see the "z" for z's, the "r-e-m" for rapid eye movement REM sleep and "roze" rhymes with doze.

<u>Miscellaneous</u>

179. Trazodone (Desyrel) *(DES-ee-rel*

tra'   = tra(y)
zo   = (hori)zo(n)
done   = (con)done

In practice, **trazodone** helps patients sleep and many students look to the "z" in trazodone to remember its' function.

III. Antidepressants

<u>Miscellaneous / SSRI</u>

180. Vilazodone (Viibryd) *(VIE-brid)*

vi   = vi(sion)
la'   = la(ser)
zo   = (hori)zo(n)
done = (con)done

Vilazodone is newer, launched in 2011, and is a selective serotonin reuptake inhibitor with partial agonism at the 5-HT_{1A} receptor.

<u>Selective serotonin reuptake inhibitors (SSRIs)</u>

181. Citalopram (Celexa) *(sell-EX-uh)*

ci   = (s)ci(ntillate)
ta'   = ta(b)
lo   = (Co)lo(rado)
pram   = pr(e)am(ble)

Most students, when seeing two drugs with the same root, **citalopram** and **escitalopram**, quickly put them into long-term memory. One student's trick was to remember that the "pram, p-r-a-m" medications are for de–p–r–ession, but "pram" is not an official stem. I associate the brand name **Celexa** with the word "relax," as some of the SSRIs are used not only for depression, but for anxiety.

182. **Escitalopram (Lexapro)** *(LECKS-uh-pro)*

es   = (m)es(s)
ci   = (s)ci(ntillate)
ta'   = ta(b)
lo   = (Co)lo(rado)
pram   = pr(e)am(ble)

While we don't have a stem for escitalopram, if you pair this medication with citalopram and these other three SSRIs, hopefully you'll remember its function. The brand name **Lexapro** takes the last four letters of **Ce<u>lexa</u>**, l-e-x-a and adds "pro, p-r-o" You can think of this as the *pro*-fessional upgrade, as **Lexapro** came after **Celexa**. It's common for an S isomer to come to market after the original drug has become available as a generic.

183. Ser<u>traline</u> (Zoloft) (ZO-loft)

ser'   = (dre)ser
tra   = (orches)tra
line   = (sa)line

One should use the "-traline," t-r-a-l-i-n-e stem to remember **ser<u>traline</u>** is an SSRI. Most students also memorize the brand name **Zoloft** first as "<u>loft</u>ing" a depressed patient's mood.

184. Flu<u>oxetine</u> *(PRO-zack)*

flu   = flu(id)
ox'   = ox
e   = (n)e(t)
tine = (sal)tine
was the first SSRI to make it to market. The "–oxetine," o-x-e-t-i-n-e ending is supposed to be for **flu<u>oxetine</u>**-like entities, but you will see "-oxetine" on the S-N-R-I antidepressant **du<u>loxetine</u> (Cymbalta)** and ADHD medication **atom<u>oxetine</u> (Strattera)**, so be careful with this particular stem.

When **flu<u>oxetine</u>** gained a new indication, for premenstrual dysphoric disorder (P-M-D-D), it also gained a new brand name: **Sarafem, pronounced and spelled** – "Sara,"s-a-r-a like the girl's name and "fem," f-e-m for feminine.

The highest ranked winged angels are Sera-p-h-i-m, so combatting P-M-D-D might be thought of as the work of angels. I don't know if that's what the drug manufacturer was going for, but that's how I've remembered it.

185. Paroxetine (brand names Paxil or Paxil C-R) *(PACKS-ill)*

pa	= (pa)pa
rox'	= (xe)rox
e	= (n)e(t)
tine	= (sal)tine

Paroxetine is similar to the SSRI **fluoxetine** with the same "oxetine" stem. **Paxil** takes "p-a-x"" from **par**oxetine. The controlled-release C-R version of **Paxil** is supposed to have fewer initial side effects and be a little easier to dose. Let's look at a related class of antidepressant by examining two S-N-R-I antidepressants.

Serotonin-Norepinephrine reuptake inhibitors (SNRIs)

186. Duloxetine (Cymbalta) *(SIM-bal-tah)*

du   = du(et)
lox'   = lox
e   = (n)e(t)
tine   = (sal)tine

While **duloxetine** ends in "oxetine," it is an SNRI, not SSRI. **Duloxetine** affects serotonin *and* norepinephrine and one can think of the "du" as duo for two. It's also pronounced like the French deux, d-e-u-x which also means two. I have never seen an unhappy cymbal player in a band and "alta" means tall in Spanish. So, I picture a happy, tall cymbal player to remember brand **Cymbalta** elevates mood.

187. **Desvenlafaxine (Pristiq)** *(prih-STEEK)*

des   = des(ks)
ven   = ven(om)
la   = (co)la
fax'   = fax
ine   = (sal)ine

We can best memorize desvenlafaxine by its stem "-faxine, f-a-x-i-n-e" as there is a related SNRI medication, **venlafaxine** as well. **Desvenlafaxine** is the enantiomer of **venlafaxine**.

188. **Venlafaxine (Effexor)** (Eff-ECKS-or)

ven   = ven(om)
la   = (co)la
fax'   = fax
ine   = (sal)ine

If you look at the "afax, a-f-a-x" in **venlafaxine** and "Effex, E-f-f-e-x" in brand name **Effexor,** you can see some commonalities as to this drug having antidepressant effects.

Tricyclic antidepressants (TCAs)

189. Ami<u>triptyline</u> (Elavil) (ELLE-uh-vill)

a	= (m)a(t)
mi	= mi(nt)
trip'	= trip
ty	= ty(pical)
line	= (sa)line

Amitriptyline's stem, is "triptyline, t-r-i-p-t-y-l-i-n-e," and I remember it's spelled t-r-*i*-p, then t-*y*-l, as this med trips up depression. Using the "tri" in **ami<u>tri</u>ptyline** helps students remember this is a "T-C-A," or *tricyclic* antidepressant. Think of the brand **Elavil** as <u>elev</u>ating the patient's mood.

190. **Doxe<u>pin</u> (Sinequan)** *(SIN-eh-quan)*

dox'	= (para)dox
e	= (n)e(t)
pin	= pin

Beating depression one student thought of as getting one's ducks in a row. She pictured "ducks, d-u-c-k-s" and duck pin, p-i-n, bowling related to dox-e-pin.

191. **Nor<u>triptyline</u> (Pamelor)** *(PAM-eh-lore)*

nor	= nor
trip'	= trip
ty	= ty(pical)
line	= (sa)line

Nortriptyline has the same -triptyline stem as amitriptyline, another TCA.

Tetracycylic antidepressant (TeCA) Noradrenergic and
specific serotonergic antidepressants (NaSSAs)

192. Mirtazapine (Remeron) *(REH-mir-on)*

mir	= (ad)mir(al)
ta'	= ta(b)
za	= (pla)za
pine	= (shi)p(mar)ine

Alternate pronunciation:

mir	= (ad)mir(al)
ta'	= ta(pe)
za	= (pla)za
pine	= (shi)p(mar)ine

I added the tricyclic antidepressants (TCAs) **doxepin** and
nortriptyline before the tetracyclic antidepressant (TeCA)
mirtazapine because three rings comes before four rings.
Remeron is a noradrenergic and specific serotonergic
antidepressant (NaSSA).

Monoamine oxidase inhibitor (MAOI)

193. Isocarboxazid (Marplan) *(MAR-plan)*

i	= "i"
so	= so(fa)
car	= car
box'	= box
a	= (comm)a
zid	= zi(ppe)d

A student came up with **Marplan** for the atypical sad man who laments, "I so carve boxes" for **isocarboxazid**.

In addition, you can take the "m" and "a" from brand **Marplan** to remember it's an M-A-O-I.

IV. Smoking Cessation

194. Bupropion (brand names Wellbutrin or Zyban) *(well-BYOO-trin, ZY-ban)*

bu   = bu(reau)
pro'   = pro(ton)
pi   = pi(ece)
on   = on

Bupropion was first marketed as brand **Wellbutrin**, an atypical antidepressant, one that didn't fit into the SSRIs, SNRIs, TCAs, or MAOIs.

Reports must have come in that patients stopped smoking while taking it, so the company repackaged the drug with a new brand name as **Zyban**, to put a "ban" on smoking.

There is some risk with this medication, especially in patients with a history of seizures.

195. Varenicline (Chantix) *(CHAN-ticks)*

va   = (lar)va,
re'   = re(d)
ni   = (k)ni(t)
cline   = cl(iff)(sal)ine

Varenicline has the "nicline, n-i-c-l-i-n-e" stem. And if you replace that "l" with an o-t, you get n-i-c-o-t-i-n-e, nicotine.

A student, in a southern drawl, remembered varenicline's generic name by saying, "With **varenicline**, 'I'm vary incline ta quit.'"

Another came up with "My chant is, 'I don't need my fix' with **Chantix**."

V. Barbiturates

196. Phenobarbital (Luminal) *(LUE-mih-nahl)*

phe	= phe(nom)
no	= no
bar'	= bar
bi	= bi(t)
tal	= tal(l), butalbital

I put the barbiturate **phenobarbital** before benzodiazepines because, chronologically, it came earlier and was a more dangerous predecessor. The barb, b-a-r-b stem classifies it as a barbiturate.

CHAPTER 5 - NEURO

VI. Benzodiazepines

197. Alprazolam (Xanax) *(ZAN-ax)*

al   = (s)al(iva)
pra'   = pra(ctice)
zo   = zo(ne)
lam   = lam(b)

Alternate pronunciation:

al   = (s)al(iva)
pra'   = pra(y)
zo   = zo(ne)
lam   = lam(b)

You should remember benzodiazepines by their two endings, "azolam, a-z-o-l-a-m" or "azepam, a-z-e-p-a-m" Be careful as a number of online and highly regarded test prep resources, including those by licensed professionals, refer to benzodiazepine stems as "pam, p-a-m" or "lam, l-a-m"

That is incorrect. This will lead you to think drugs that are not benzos are in the class. For example, vera*pam*il **(Calan)** is a calcium channel blocker used for hypertension and *lam*otrigine **(Lamictal)** is an antiepileptic.

I've even seen a well-regarded resource indicate that **citalopram** has a "pam" suffix at the end, and that's understandable as some patients drop the "r," pronouncing it "citalo-pam," but this is incorrect. Citalopram is an SSRI with antianxiety properties, but not a benzo.

A student came up with this rhyme to connect the generic and brand name.

Alprazolam has one z; benzodiazepine has two; **Xanax** sounds like a "z" to help you get a snooze and "x's" out anxiety too.

198. Midazolam (Versed) *(VER-said)*

mi	= mi(nt)
da'	= da(y)
zo	= zo(ne)
lam	= lam(b)

You should remember **midazolam** through the "azolam, a-z-o-l-a-m" stem, but there are other tips you can use.

There are two m's in **midazolam** for the memories you can't form since **midazolam** causes anterograde amnesia.

Just as your *ante*brachium is your forearm, and the *ante* is the money poker players put out before the dealer deals, *ante*rograde amnesia is the inability to form future memories. Alternatively, you can use the brand name; I can't remember the "verse you just said" for brand **Versed**.

199. Clonazepam (Klonopin) *(KLON-uh-pin)*

clo	= clo(ver)
na'	= na(vy)
ze	= ze(d)
pam	= pam(phlet)

Clonazepam should be remembered from the "azepam, a-z-e-p-a-m" stem.

The brand name **Klonopin** uses the phonetic spelling of generic **clonazepam's** first four letters "c-l-o-n."

200. Lorazepam (Ativan) *(AT-eh-van)*

lo	= (Co)lo(rado)
ra'	= ra(y)
ze	= ze(d)
pam	= pam(phlet)

Remember **lorazepam** through the "–azepam" stem or think about the brand name **Ati-*van* van**quishing anxiety.

201. Tem<u>azepam</u> (Restoril) *(REH-stoh-rill)*

te   = te(nt)
ma'   = ma(ze)
ze   = ze(d)
pam   = pam(phlet)

Alternate pronunciation:

te   = te(nt)
ma'   = ma(p)
ze   = (ha)ze(l)
pam   = pam(phlet)

Temazepam is a benzodiazepine marketed for sleep disorders, i.e. for "<u>rest</u>" or "<u>resto</u>ration" as its brand name Restoril implies.

VII. Non-benzodiazepine / non-barbiturate

202. Buspirone (Buspar) *(BYOU-spar)*

bu'   = bu(reau)
spi   = spi(ll)
rone   = (d)rone

As nonfiction literature is "not" fiction, we classify **buspirone** as a non-barbiturate, non–benzodiazepine.

VIII. ADHD medications

<u>Stimulant – Schedule II</u>

203. Amphetamine/Dextroamphetamine (Adderall) *(ADD-er-all)*

am   = (l)am(p)
phe' = phe(nomenon)
ta   = (da)ta
mine = (hista)mine

dex = dex(terity)
tro = (me)tro
am   = (l)am(p)
phe' = phe(nomenon)
ta   = (da)ta
mine = (hista)mine

The **Adderall** brand name reminds us that once a student can focus, it's no problem to add 'er all up in math class.

204. Dexmethylphenidate (Focalin) *(FOE-ca-lin)*

dex   = dex(terity)
meth = meth(od)
yl   = (s)yl(lable)
phe' = phe(nomenon)
ni   = (k)ni(t)
date = date

When I was a student, I could remember that **Focalin** and **Concerta** both had **methylphenidate** in their names, but I could never remember which was **dexmethylphenidate** and which was **methylphenidate**.

Then I thought of the "F" in "**Focalin**" as following the "d-e" alphabetically in **dexmethylphenidate** to help me.

To remember **Focalin** is for ADHD, you think of **Focalin** helping a patient focus.

205. Lisdexam<u>fetamine</u> (Vyvanse) (*VIE-vanse*)

lis	= lis(t)
dex	= dex(terity)
am	= (l)am(p)
fe'	= fe(n)
ta	= (da)ta
mine	= (hista)mine

206. Methylphenidate (brand names Ritalin and Concerta) *(con-CERT-uh)*

meth	= meth(od)
yl	= (s)yl(lable)
phe'	= phe(nomenon)
ni	= (k)ni(t)
date	= date

Methylphenidate has many brand names, including **Ritalin** and **Concerta.** Most students seem to already know **methylphenidate,** but remember that **Concerta** can help a patient <u>concentrate</u>, taking c-o-n-c-e-r and t from concentrate; **Concerta**.

To remember it's an amphetamine, one student said she thought of staying up all night at a <u>concert</u> with **Concerta**. **Concerta** is a long-acting medication that needs to be taken only once a day.

I alphabetized the ADHD medications: **Amphetamine** with **dextroamphetamine (Adderall)** and **Lisdexam<u>fetamine</u> (Vyvanse),** which has a British "–fetamine" ending.

Non-stimulant – non-scheduled

207. Atom<u>oxetine</u> (Strattera) *(stra-TER-uh)*

a   = "a"
to   = (pis)to(n)
mox' = mox(ie)
e   = (n)e(t)
tine   = (sal)tine

Note the "oxetine" stem here is not an SSRI antidepressant, but a non-stimulant medication for ADHD.

The brand name **Strattera** uses the s-t-r from <u>str</u>aighten and a-t-t-e from <u>atte</u>ntion, so straighten a patient's attention, **Strattera**.

IX. Bipolar Disorder

Simple salt

208. Lithium (Lithobid) *(LITH-oh-bid)*

li'   = li(ly)
thi   = thi(ef)
um   = (g)um

Lithium sits in the same group on the far left of the Periodic Table of Elements as the Latin *Natrium* (N-a), commonly known as the chemical element sodium.

The body has trouble telling the difference between lithium and sodium, and too much or too little sodium can wreak havoc on **lithium** levels.

A way to remember this is the saying, "Where the salt goeth, the **lithium** goeth."

The "b-i-d" in the brand name **Lithobid** is the Latin *bis-in-die*, or twice-daily for the dosing schedule.

XI. Schizophrenia

First generation antipsychotic (FGA) (low potency)

209. Chlorpromazine (Thorazine) *(THOR-uh-zeen)*

chlor = chlor(ine)
pro'   = pro(ton)
ma   = (paja)ma
zine   = (maga)zine

Chlorpromazine was the first antipsychotic. While it carries side effects, it represented a new treatment option for schizophrenic patients – a 1st generation.

Generational classifications are especially important in antipsychotics because of differences in both side effects and effects on positive versus negative symptoms.

Thor is a mythical god, and you can think of brand **Thorazine** as helping people who have delusions of mythical people.

<u>First generation antipsychotic (FGA) (high potency)</u>

210. Halo<u>peridol</u> (Haldol) *(HAL-doll)*

ha	= ha(t)
lo	= (Co)lo(rado)
pe'	= pe(n)
ri	= ri(nse)
dol	= dol(l)

Use the "-peridol, p-e-r-i-d-o-l" stem to recognize this 1st-generation high potency drug.

Many students think of the "halo" in **haloperidol** as the halo that sits *high* on an angel's head to remember this is *high* potency.

To make the brand name **Haldol**, they just took the first and last three letters of the generic name **hal**operi**dol**; **Haldol**.

Second-generation antipsychotic (SGA)

I added two second-generation antipsychotics, **aripiprazole (Abilify)** and **olanzapine (Zyprexa)** alphabetically.

211. Aripiprazole (Abilify) *(ah-BILL-ih-fie)*

a   = a(ir)
ri   = ri(nse)
pi'   = pi(n)
pra   = (su)pra
zole   = (fe)z(p)ole

The WHO, world health organization discourages the "–piprazole" stem because it has the PPI "–prazole" stem in it. **Abilify** helps a schizophrenic have more "ability to function in society."

212. Olanzapine (Zyprexa) *(zie-PRECKS-ah)*

o   = "o"
lan'   = lan(ce)
za   = (pla)za
pine   = (shi)p(mar)ine

213. Risperidone (Risperdal) *(RIS-per-doll)*

ri   = ri(nse)
spe'   = spe(ck)
ri   = ri(nse)
done   = (con)done

Note the stem "–peridol, p-e-r-i-d-o-l" from **haloperidol** and "–peridone, p-e-r-i-d-o-n-e" from **risperidone** are similar; that can help you remember both of these are antipsychotics.

The brand name **Risperdal** and generic name **risperidone** share the opening letters "r-i-s-p-e-r, "risper. One student thought of risper and whisper, as in calming the whispering voices.

214. Que<u>tiapine</u> (Seroquel) *(SEAR-uh-kwell)*

que	= que(ll)
ti'	= ti(e)
a	= (comm)a
pine	= (shi)p(mar)ine

In **quetiapine**, you should use the "-tiapine, t-i-a-p-i-n-e" stem to recognize that this is a 2nd-generation antipsychotic.

If you switch the "i" and "t" in **quetiapine**, q-u-e-t-i-a-p-i-n-e, you get the word "quiet," as in quieting the voices.

Seroquel, the brand name, shares the "q-u-e" from **que**tiapine and <u>quel</u>l means to silence someone.

XII. Antiepileptics

Traditional antiepileptics

215. Carbamazepine (Tegretol) *(TEG-reh-tawl)*

car = car
ba = (tu)ba
ma' = ma(ze)
ze = ze(d)
pine = (shi)p(mar)ine

While **carbamazepine's** "-pine," [[pronounced, "peen,"]] p-i-n-e is the stem, it's not very useful because the –"-pine" ending just means a chemical has three rings.

Instead, think either of seizures being "carbed" (curbed) or of being "amazed" that the seizures are "sto-pping" using the letters inside **carb-amaze-pine**.

I remember the brand name **Tegretol** has the scattered letters "t-r-o-l" in con<u>trol</u>, as in to control seizures; **Tegretol**.

216. Divalproex (Depakote) *(DEP-uh-coat)*

di = di(ce)
val' = val(ley)
pro = pro(ton)
ex = ex(po) [Note: not "egg zit"]

While you can find "v-a-l" in many medication names, it is helpful to think of the "val" in di<u>val</u>proex and it's similarity with the "vul, v-u-l" in con<u>vul</u>sions. I know it's a stretch.

One student thought of **dival-pro-ex** as a <u>pro</u> at <u>ex</u>tracting seizures; **divalproex**.

217. Pheny<u>toin</u> (Dilantin) *(DYE-lan-tin)*

phe = phe(nomenon)
ny' = (po)ny
to = (toe)
in = in, (t)in, polymyxin B

The "toin, t-o-i-n" stem helps you remember **phenytoin** is an antiepileptic.

Dilantin and shakin' almost rhyme.

Newer antiepileptics

218. Gabapentin (Neurontin) *(NER-on-tin)*

ga' = ga(s)
ba   = (tu)ba
pen = pen
tin = tin

The "gab, g-a-b" stem is a little misleading. Neither **gabapentin** nor **pregabalin** directly affect gamma-amino-butyric-acid (GABA) [[gah-bah]] receptors. However, having them both have the same stem helps set a memory device for the newer antiepileptics.

The "neu, n-e-u" [[pronounce "new"] in **Neurontin** is one way to remember that this is a newer drug.

219. Lamotrigine (Lamictal) *(lam-ICK-tal)*

la   = (co)la
mo' = mo(at)
tri = tri(p)
gine = (aspara)gine

Newer antiepileptics include **lamotrigine** and the brand name **Lamictal** has "ictal," in it, meaning seizure.

220. Levetiracetam (Keppra) *(KEP-rah)*

le   = le(tter)
ve   = ve(teran)
ti   = ti(ara)
ra' = (au)ra
ce   = ce(iling)
tam = tam(ale)

In the brand Keppra, you can use the K-e-p, Kep and p-r-a, pra to think that an epileptic "kept praying" the seizures would stop.

221. Oxcarbazepine (Trileptal) *(try-LEP-tahl)*

ox'	= ox
car	= car
ba'	= (bay)
ze	= ze(d)
pine	= (shi)p(mar)ine

By changing the "carb" to "curb" you can think of curbing seizures with **oxcarbazepine**. The brand **Trileptal** has the e-p-i-l from epileptic.

222. Topiramate (Topamax) *(TOE-pah-max)*

to	= to(e)
pi'	= pi(ece)
ra	= (au)ra
mate	= mate

In both the generic and brand names, topiramate and Topamax, you can use the t-o-p as in s-t-o-p, stop seizures.

223. Ziprasidone (Geodon) *(GEE-oh-don)*

zi	= zi(pper)
pra'	= pra(y)
si	= si(t)
done	= (con)done

One student said that seeing **ziprasidone** has p-r-a for "pray" then i-s backwards and d-o-n-e, done for "pray this seizure is done." The word "done" can also be found in the the brand **Geodon**.

224. Pregabalin (Lyrica) *(LEER-eh-ca)*

pre   = pre(mie)
ga'   = ga(s)
ba   = (tu)ba
lin   = lin(t)

A <u>lyre</u> is a musical instrument and a <u>lyric</u> is a line in a song.

Either way, you can think of a seizure coming back into harmony with **Lyrica**.

XIII. Parkinson's, Alzheimer's, Motion sickness

<u>Parkinson's</u>

225. Benz<u>tropine</u> mesylate (Cogentin) *(co-JEN-tin*

benz	= (Mercedes)Benz
tro'	= (me)tro
pine	= (shi)p(mar)ine

The Parkinson's medication **benztropine mesylate (Cogentin)** hints at <u>cognition</u>.

226. Levo<u>dopa</u> with Carbi<u>dopa</u> (Sinemet) *(SIN-uh-met)*

le	= le(tter)
vo	= (di)vo(t)
do'	= do(ugh)
pa	= (pa)pa
car	= car
bi	= bi(t)
do'	= do(ugh)
pa	= (pa)pa

The stem dopa, d-o-p-a in both levodopa and carbidopa help remind us that increased dopamine is critical to dealing with the disease. The brand *Sin*emet combines **levodopa** and **carbidopa** to work *syn*ergistically.

Carbidopa doesn't actually have an antiparkinsonian effect, but it reduces the breakdown of **levodopa** so more is available to the patient from a smaller dose.

227. **Selegiline (Eldepryl)** *(EL-duh-pril)*

se   = se(t)
le'   = le(tter)
gi   = (ma)gi(c)
line   = (sa)line

The "–giline," g-i-l-i-n-e stem should be your hint that **selegiline** is a Parkinson's medication.

You can find the letters of the word senile in the generic name **selegiline**. I used to confuse this as an Alzheimer's medication, but it's not, so I was being a little senile.

The brand name **Eldepryl** helps you remember that it relieves symptoms of Parkinson's disease, a condition more prevalent in the "elderly;" **Eldepryl**.

228. Pramipexole (Mirapex ER) *(MEER-ah-pecks ee-are)*

pra   = pra(ctice)
mi   = mi(nt)
pex'   = (a)pex
ole   = (m)ole

The brand name **Mirapex** twists around the generic name pramipexole.

229. Ropinirole (Requip, Requip XL) *(REE-kwip, REE-kqip ex-ell)*

ro   = ro(w)
pi'   = pi(n)
ni   = (k)ni(t)
role   = role

"Equips" a patient to deal with Parkinson's.

Alzheimer's

230. Memantine (Namenda) *(Nuh-MEN-duh)*

me   = me(n)
man' = man
tine   = (sal)tine

The generic name **memantine** has "mem, m-e-m" for memory in it.

The brand name **Namenda** comes from <u>N</u>-<u>M</u>ethyl-<u>D</u>-<u>a</u>spartate **(N-M-D-A)**, the receptor it antagonizes.

One of my students always thought of **Namenda** as <u>mend</u>ing the brain; **Namenda**.

231. Donepezil (Aricept) *(AIR-eh-sept)*

do   = do(t)
ne'   = ne(t)
pe   = pe(n)
zil   = (Bra)zil

When I think of **donepezil** as an Alzheimer's medication, I think, "I <u>don</u>'t remember <u>zil</u>ch!" taking the "d-o-n" from the front of donepezil and the "z-i-l" from the back. The brand name **Aricept** improves per<u>cept</u>ion and Alzheimer's patients' powers of recollection.

Motion sickness

232. Scopolamine (Transderm-Scop) *(trans-DERM SCOPE)*

sco   = sco(pe)
po'   = po(t)
la   = la(wn)
mine = (hista)mine

Transderm-Scop is a transdermal form of **scopolamine**, for motion sickness. "Trans" means across, "derm" means "skin," so across-the-skin **scopolamine**; **Transderm-Scop**

Chapter 6 - Cardio

I. OTC Antihyperlipidemics and antiplatelet

<u>Antihyperlipidemics</u>

233. **Omega-3-acid ethyl esters (Lovaza)** *(Loh-VAH-zah)*

Omega-3-fatty acids are available over-the-counter, but there is also a prescription version that undergoes rigorous FDA testing.

Lovaza is a prescription brand name, but you can find **omega-3-fatty acids** over-the-counter commonly labeled as "Fish Oil."

234. **Niacin (Niaspan)** *(NYE-uh-span)*

ni'   = ni(ght)
a   = (comm)a
cin   = cin(der)

A vitamin like **niacin** can reduce cholesterol levels in the body, however it may cause facial flushing that an **aspirin** thirty minutes before treatment prevents.

This flushing unfortunately gets in the way of helping many patients who would otherwise benefit from an inexpensive cholesterol lowering medication.

Antiplatelet

235. Aspirin (Ecotrin) *(ECK-oh-trin)*

as' = as(p)
pi = pi(n)
rin = rin(se) [Note: not r-i-n-e]

The 81mg daily **aspirin** dosage is not for analgesia or fever reduction. Rather, it keeps platelets from sticking, helping prevent strokes and heart attacks.

II. Diuretics

<u>Osmotic</u>

236. Mannitol (Osmitrol) *(OZ-meh-trawl)*

man' = man
ni   = (k)ni(t)
tol   = tol(erate)

Mannitol, an osmotic diuretic reduces intracranial pressure in an emergency.

The brand name **Osmitrol** combines the class of medication "<u>osm</u>o<u>t</u>ic," and adds that it helps con<u>trol</u> brain swelling, taking the o-s-m-i from osmotic and t-r-o-l from control; **Osmitrol**.

The actor Bruce Lee died from this event.

<u>Loop</u>

237. Furo<u>se</u>mide (Lasix) *(LAY-six)*

fu   = fu(mes)
ro'   = ro(w)
se   = se(t)
mide   = (gu)m(sl)ide

Chemists named this class of diuretics after the part of the nephron the drug works in, the Loop of Henle.

While the "semide, s-e-m-i-d-e" stem indicates a "furosemide-type" diuretic, that's like defining a word with the word itself.

One student said, "I have to pee <u>fur</u>iously" as her mnemonic since loop diuretics produce significant diuresis taking the f-u-and r from furosemide.

The brand name **Lasix** indicates it <u>las</u>ts <u>six</u> hours; **Lasix**.

Thiazide

238. Hydrochloro<u>thiazide</u> (Microzide) *(MY-crow-zide)*

hy	= hy(brid)
dro	= dro(ne)
chlor	= chlor(ine)
o	= "o"
thi'	= thi(gh)
a	= (comm)a
zide	= z(ipper)(r)ide

Thiazide diuretics get their class name from the stem –
thiazide, t-h-i-a-z-i-d-e of generic drugs like
hydrochlorothiazide.

The abbreviation H-C-T-Z comes from "h" for <u>h</u>ydro, "c" for
<u>c</u>hloro, "t" for <u>t</u>hia, and "z" for <u>z</u>ide.

Thiazides don't produce as much diuresis as loop diuretics,
but are excellent for initial treatment of hypertension.

While the "hydro" in **hydrochlorothiazide** stands for the
<u>hydro</u>gen atom, you can think of "hydro" as "water" for
diuretic.

The brand **Microzide** has thia<u>zide</u>'s last four letters, z-i-d-e,
Microzide.

Potassium sparing and thiazide

239. Triamterene with Hydrochloro<u>thiazide</u> (Dyazide) *(DIE-uh-zyde)*

tri	= tri(angle)
am'	= (l)am(p)
te	= te(nt)
rene	= (se)rene
hy	= hy(brid)
dro	= dro(ne)
chlor	= chlor(ine)
o	= "o"
thi'	= thi(gh)
a	= (comm)a
zide	= z(ipper)(r)ide

The combination of a potassium sparing diuretic (**triamterene**) and thiazide diuretic (**hydrochlorothiazide**) keeps potassium levels in balance while producing modest diuresis.

The brand name **Dyazide** takes the first two letters of triamterene's brand name **Dyrenium** plus the last five letters of **hydrochloro<u>thiazide</u>**, so d-y plus i-a-z-i-d-e; **Dyazide**.

Potassium sparing

240. Spironolactone (Aldactone) *(Al-DAK-tone)*

spi	= spi(ll)
ro	= ro(w)
no	= no
lac'	= (li)lac
tone	= tone

Spironolactone is another potassium sparing diuretic like triamterene, but this drug can cause gynecomastia.

Gynecomastia is an enlargement of male breasts. To remember **spironolactone** works in the collecting duct, I look at the "lactone," and think "last one," taking the l-a and o-n-e from lactone.

I know a lactone is a kind of chemical structure, but its place of action sticks in my head with this mnemonic.

To come up with the brand name, the manufacturer simply replaced the "spironol" of **spironolactone** with "ald, a-l-d" to make **Aldactone**.

The "ald" is especially important because **spironolactone** blocks *ald*osterone, an important steroid hormone that helps the body retain sodium and water when blood pressure drops.

Electrolyte replenishment

241. Potassium chloride (K-Dur) *(Kay-Dur)*

po = po(tato)
tas' = tas(k)
si = (enthu)si(astic)
um = (g)um

chlor' = chlor(ine)
ide = (sl)ide

Potassium chloride is a supplement often administered when a potassium sparing diuretic is contraindicated or when a loop diuretic lowers a patient's potassium levels.

The "K" in **K-Dur** is the chemical symbol for potassium.

The "Dur, d-u-r," is for long _dur_ation.

III. Understanding the Alphas and Betas

Alpha-1 antagonist

242. Doxazosin (Cardura) *(car-DUR-uh)*

dox   = (para)dox
a'   = (m)a(t)
zo   = zo(ne)
sin   = (ba)sin

Doxazosin's blockade of alpha-1 receptors causes vasodilation and subsequent reduction in blood pressure. Memorize the stem "azosin, a-z-o-s-i-n" as an alpha-blocker.

You can also link the brand name, as **Cardura** provides durable cardiac relief of hypertension, using the d-u-r-a from durable and c-a-r from cardiac to form **Cardura**.

243. Terazosin (Hytrin) *(HIGH-trin)*

ter   = ter(se)
ra'   = ra(ttle)
zo   = zo(ne)
sin   = (ba)sin

The cardio drugs are mostly combinations of old drugs. The alpha-1 antagonist **terazosin** has the "–azosin" ending and **Hytrin** takes six letters from "hypertension."

Alpha-2 agonist

244. Clonidine (Catapres) *(CAT-uh-press)*

clo' = clo(th)
ni = (k)ni(t)
dine = (bir)di(e)(tu)ne

Clonidine works in the brain by affecting alpha-*two* receptors
to reduce peripheral vascular resistance. You can look at the
brand name **Catapres** and think of <u>cata</u>bolize (break down)
<u>pres</u>sure (blood pressure) using the c-a-t-a from catabolize
and p-r-e-s from pressure; **Catapres**.

Prescribers also use **clonidine** in A-D-H-D as single therapy
or with stimulants like **methylphenidate,** brand name
Concerta. I had the weirdest experience at the gym where a
parent had a loud and lengthy discussion with a psychiatrist
about her child's **clonidine** and **Concerta** while lifting
weights, they were talking as loud as a rock *concert*.

I never forgot the **clonidine and Concerta** tandem after that.

Beta-blocker – 1st-generation – non-beta selective

245. Propran<u>olol</u> (Inderal) *(IN-dur-all)*

pro	= pro(ton)
pra'	= pra(nce)
no	= (can)no(n)
lol	= lol(lipop)

Propranolol's last four letters have the "–olol, [[uh-lawl]] o-l-o-l" beta-blocker stem. The "o-l-o-l" looks like two letter "b's" backwards to remember (b) beta (b) blocker.

Alternatively, if you "<u>o</u>h, <u>l</u>augh <u>o</u>ut <u>l</u>oud," using the o-l-o-l from those four words, you can think of your heart rate going down from stress relief.

If you think of the last "al," a-l in the brand name **Inder<u>al</u>** as it blocks "all" beta-receptors, you can remember this is a non-selective beta-adrenergic antagonist.

Beta-blockers – 2nd-generation – beta selective

246. Aten<u>olol</u> (Tenormin) *(Teh-NOR-min)*

a	= (comm)a
te'	= te(nt)
no	= (can)no(n)
lol	= lol(lipop)

While the "olol, o-l-o-l" in **atenolol** identifies this medication as a beta-blocker, a student *does* have to memorize that **atenolol** is 2nd generation.

She can do that by memorizing atenolol's position *after* a non-selective first generation **propranolol** in this book *or* by seeing the "ten" in **aten<u>olol</u>** and knowing it's divisible by *two*.

The "Ten, T-e-n" in the brand name **Tenormin** also matches to the "ten, t-e-n" in **aten<u>olol</u>** to help connect the brand and generic names *Ten*ormin and a*ten*olol.

247. Aten<u>olol</u> / Chlorthalidone (Tenoretic) *(ten-or-EH-tick)*

a	= (comm)a
te'	= te(nt)
no	= (can)no(n)
lol	= lol(lipop)
chlor	= chlor(ine)
tha'	= tha(tch)
li	= li(ly)
done	= (con)done

Atenolol with **chlorthalidone** is **Tenoretic**. **Tenoretic** takes the "T-e-n-o-r" from **Tenormin** and adds "r-e-t-i-c" from "diu<u>retic.</u>

248. Bisopr<u>olol</u> / Hydrochloro<u>thiazide</u> (Ziac) *(ZIE-ack)*

bi	= bi(t)
so'	= so(fa)
pro	= (a)pro(n)
lol	= lol(lipop)
hy	= hy(brid)
dro	= dro(ne)
chlor	= chlor(ine)
o	= "o"
thi'	= thi(gh)
a	= (comm)a
zide	= z(ipper)(r)ide

Note the beta-blocker stem –olol in bisoprolol and thiazide diuretic stem –thiazide in hydrochlorothiazide to recognize the ingredients in the combination medicine. Combination medicines from different drug classes help reduce adverse effects by having lower doses of two drugs rather than a higher dose of one.

249. **Metoprolol tartrate (Lopressor)** *(low-PRESS-or)*

me	= me(n)
to'	= to(e)
pro	= (a)pro(n)
lol	= lol(lipop)

suc'	= suc(ces)
ci	= (s)ci(ntillate)
nate	= (in)nate

Practitioners rarely highlight the distinction in salts like tartrate and succinate, but it's important to recognize, as **metoprolol** *tartrate* and **metoprolol** *succinate* work for different lengths of time.

It would have been nice if the succinate was short acting and the tartrate, long acting; then alphabetical order would have worked or "s" for short acting.

That it goes contrary to this logic is how I personally remember which is which.

You can use the brand name **Lopressor** to remind you that **Lopressor** lowers blood pressure, using the l-o from lowers and p-r-e-s-s from pressure; Lopressor.

250. Metoprolol succinate (Toprol X-L) *(TOE-prall ex-ell)*

me	= me(n)
to'	= to(e)
pro	= (a)pro(n)
lol	= lol(lipop)

tar'	= tar
trate	= (concen)trate

Metoprolol succinate is a long-acting form of **metoprolol**. The X-L, often used to identify clothing as extra-large, indicates an extra-long acting effect in medications like **Toprol XL**.

Beta-blocker – 3rd-generation – non-beta selective
vasodilating

251. Carve<u>dil</u>ol (Coreg) *(CO-reg)*

car'   = car
ve   = ve(teran)
di   = di(nner)
lol   = lol(lipop)

I'm not sure if it was intentional to create a kind of hybrid
stem with the "dil, d-i-l" replacing the first "o" in "olol, o-l-o-
l," but you can remember **carvedilol** works by both
vasodilation and beta-blockade in this way. The only official
stem, however, is the "dil, d-i-l."

I remember the brand name **Coreg** because it <u>reg</u>ulates
<u>co</u>ronary function, using the r-e-g from regulates and c-o
from coronary; **Coreg.**

252. Labe<u>tal</u>ol (Normodyne) *(NOR-moe-dine)*

la   = la(b)
be'   = be(ta)
ta   = (da)ta
lol   = lol(lipop)

has "<u>beta</u>" in the generic name. Instead of "–olol" for beta-
blocker, the stem is –alol for alpha / beta-blocker.

157

253. Nebiv<u>olol</u> (Bystolic) *(bye-STAHL-ick)*

ne   = ne(t)
bi' = bi(t)
vo   = (di)vo(t)
lol = lol(lipop)

Nebivolol's brand name **Bystolic** takes letters from sy<u>stolic</u> (the top blood pressure number) and dia<u>stolic</u> (the bottom number).

IV. Renin-angiotensin-aldosterone system (RAAS)

I grouped renin angiotensin aldosterone system (RAAS) drugs in alphabetical order. For drugs that have a second component, I put the single drug first. I won't go into the medications too much because by learning "–pril," "-sartan," and "-thiazide," you will know what each drug from 254 to 270 is for.

ACE Inhibitors (ACEIs)

254. Benazepril / HCTZ (Lotensin HCT) *(loe-TEN-SIN)*

be	= be(d)
na'	= na(p)
ze	= ze(d)
pril	= pril(l)

hy	= hy(brid)
dro	= dro(ne)
chlor	= chlor(ine)
o	= "o"
thi'	= thi(gh)
a	= (comm)a
zide	= z(ipper)(r)ide

Benazepril with the ACE inhibitor stem **-pril** becomes **Lotensin HCT** when the manufacturer adds **hydrochlorothiazide**, a thiazide diuretic. The "tensin" in the brand name **Lotensin** is for hypertension.

255. Enala<u>pril</u> (Vasotec) *(VA-zo-teck)*

e	= (n)e(t)
na'	= na(p)
la	= (co)la
pril	= pril(l)

Sometimes students simply refer to the ACE inhibitors like **enalapril** as "prils" based on the stem "pril, p-r-i-l." While patients take **enalapril** orally, **enalaprilat** with the ending p-r-i-l-a-t, is an injectable form and active metabolite of **enalapril**.

The brand **Vasotec** alludes to <u>vas</u>odilation on the <u>vas</u>culature, using the v-a-s in Vasotec.

256. Fosinopril (Monopril) *(MON-oh-pril)*

fo	= fo(e)
si'	= si(t)
no	= no
pril	= pril(l)

is unusual in that it has the ACE inhibitor stem –pril in both the brand and generic names.

257. Quina<u>pril</u> (Accu<u>pril</u>) *(ACK-you-pril)*

qui'	= qui(nce)
na	= (bana)na
pril	= pril(l)

is also unusual in that it has the ACE inhibitor stem –pril in both the brand and generic names.

Brand names don't have stems, but the drug company decided to put those letters in for easy recognition, I believe.

258. **Lisinopril (Zestril)** *(ZES-tril)*

li   = li(ght)
si'   = si(t)
no   = no
pril   = pril(l)

Alternate pronunciation:

li   = li(ly)
si'   = si(t)
no   = no
pril   = pril(l)

Lisinopril, like **enalapril**, has the "pril, p-r-i-l" stem and works to block the vasoconstricting effects of angiotensin II.

A student came up with "**Lisinopril** thrills an overworked heart, blocking angiotensin II from getting a start."

259. Lisinopril / Hydrochlorothiazide (Zestoretic) *(Zess-TORE-eh-tick)*

li   = li(ght)
si'   = si(t)
no   = no
pril   = pril(l)

hy   = hy(brid)
dro   = dro(ne)
chlor   = chlor(ine)
o   = "o"
thi'   = thi(gh)
a   = (comm)a
zide   = z(ipper)(r)ide

When a manufacturer adds **hydrochlorothiazide** to **lisinopril**, it becomes brand **Zestoretic** by adding the last letters of "diuretic" to the name.

260. Rami<u>pril</u> (Altace) *(ALL-tayse)*

ra   = ra(ttle)
mi'   = mi(nt)
pril   = pril(l)

The brand **Altace** cleverly has the "ace" from ACE inhibitor in the name as in it will alter the ACE enzyme.

Angiotensin II receptor blockers (ARBs)

261. Candesartan (Atacand)*(AT-ah-kand)*

```
can    = can
de     = de(sk)
sar'   = sar(dine)
tan    = tan
```

Learn angiotensin II receptor blockers by the suffix "sartan, s-a-r-t-a-n," you'll hear this repeated over the next few drug names.

262. Irbesartan (Avapro) *(AVE-ah-pro)*

```
ir     = (f)ir
be     = be(d)
sar'   = sar(dine)
tan    = tan
```

263. Irbesartan / Hydrochlorothiazide (Avalide) *(AV-ah-lide)*

```
ir     = (f)ir
be     = be(d)
sar'   = sar(dine)
tan    = tan

hy     = hy(brid)
dro    = dro(ne)
chlor  = chlor(ine)
o      = "o"
thi'   = thi(gh)
a      = (comm)a
zide   = z(ipper)(r)ide
```

264. Losartan (Cozaar) *(CO-zar)*

```
lo     = (Co)lo(rado)
sar'   = sar(dine)
```

tan   = tan

The brand name **Cozaar** looks like it has R-A-A-S backwards (for <u>r</u>enin-<u>a</u>ngiotensin-<u>a</u>ldosterone-<u>s</u>ystem) with a "z" replacing the "s."

265. Lo<u>sartan</u> / Hydrochloro<u>thiazide</u> (Hyzaar) *(HIGH-zahr)*

lo   = (Co)lo(rado)
sar'   = sar(dine)
tan   = tan

hy   = hy(brid)
dro   = dro(ne)
chlor   = chlor(ine)
o   = "o"
thi'   = thi(gh)
a   = (comm)a
zide   = z(ipper)(r)ide

266. Olme<u>sartan</u> (Benicar) *(BEN-eh-car)*

ol   = ol(d)
me   = me(n)
sar'   = sar(dine)
tan   = tan

Olmesartan is another ARB identified by its "sartan, s-a-r-t-a-n" stem.

The brand name **Benicar** hints that the drug will <u>ben</u>efit the <u>car</u>diac system, taking the b-e-n from benefit and c-a-r- from cardiac to make **Benicar**.

267. Olme<u>sartan</u> / HCTZ (Benicar HCT) *(BEN-ih-car aitch-see-tee)*

ol   = ol(d)
me   = me(n)
sar'   = sar(dine)

tan   = tan

hy   = hy(brid)
dro   = dro(ne)
chlor   = chlor(ine)
o   = "o"
thi'   = thi(gh)
a   = (comm)a
zide   = z(ipper)(r)ide

268. Telmisartan / HCTZ (Micardis HCT) *(my-CAR-diss aitch-see-tee)*

tel   = tel(ephone)
mi   = mi(nt)
sar'   = sar(dine)
tan   = tan

hy   = hy(brid)
dro   = dro(ne)
chlor   = chlor(ine)
o   = "o"
thi'   = thi(gh)
a   = (comm)a
zide   = z(ipper)(r)ide

269. Valsartan (Diovan) *(DYE-oh-van)*

val   = val(ley)
sar'   = sar(dine)
tan   = tan

Identify the generic name **valsartan** by its "sartan, s-a-r-t-a-n" stem. **Diovan** has three of the letters of the generic name valsartan, the v-a-n to help connect the two.

165

270. Valsartan / HCTZ (Diovan HCT) *(DYE-oh-van)*

val	= val(ley)
sar'	= sar(dine)
tan	= tan

hy	= hy(brid)
dro	= dro(ne)
chlor	= chlor(ine)
o	= "o"
thi'	= thi(gh)
a	= (comm)a
zide	= z(ipper)(r)ide

The manufacturers of ARBs similarly create brand names for new combination products by adding either "-lide," "Hy-," or "HCT." You can easily recognize the calcium channel blocker additions with the "-dipine" for the CCB dihydropyridine class and "-vastatin" for the HMG-CoAs, the statins, and "–pril" for the A-C-E-Is.

V. Calcium channel blockers (CCBs)

Non-dihydropyridines

271. Diltiazem (Cardizem) *(CAR-deh-zem)*

dil = dil(l)
ti' = ti(e)
a = (comm)a
zem = (ec)zem(a)

The "–tiazem, t-i-a-z-e-m" stem identifies **diltiazem** as a non-dihydropyridine.

The brand name **Cardizem** adds the first five letters from cardiac, c-a-r-d-i to the last three letters of the generic diltiazem, z-e-m, to make **Cardizem**.

272. Verapamil (Calan) *(KALE-en)*

ve = ve(rb)
ra' = ra(ttle)
pa = (pa)pa
mil = mil(k)

Verapamil has the "pamil, p-a-m-i-l" stem. One of my students came up with "Vera and Pam are ill and need this calcium blocking cardiac pill," adding the V-e-r-a from Vera and P-a-m from Pam to make **verapamil**.

Often we associate **verapamil** with constipation. My grandmother, a Navy nurse, used to put a **verapamil** tablet on my grandfather's breakfast cereal spoon.

I always thought my grandfather was silently praying before he ate. When I finally asked him why he was so quiet, he said something to the effect of, "I'm deciding whether I want to eat or poop today."

The brand name **Calan** takes c-a-l from the word calcium, and a-n from channel blocker, make C-a-l-a-n, **Calan**.

167

Dihydropyridines

273. Amlodipine (Norvasc) *(NOR-vasc)*

am   = (l)am(p)
lo'   = (Co)lo(rado)
di   = di(nner)
pine = (shi)p(mar)ine

Students usually recognize **amlodipine's** "dipine, d-i-p-i-n-e" stem, not only as a dihydropyridine, taking the d-i-p-i-n-e from the chemical class, but also as a dip, d-i-p, in, i-n blood pressure; **amlodipine**.

A way to remember the brand name **Norvasc** is to think of the "n-o-r" from normalizes and the "v-a-s-c" from vasculature; **Norvasc**.

274. Amlodipine/ Atorvastatin (Caduet) *(CAH-due-et)*

am   = (l)am(p)
lo'   = (Co)lo(rado)
di   = di(nner)
pine = (shi)p(mar)ine

a   = (comm)a
tor' = tor(so)
va   = (lar)va
sta = sta(ff)
tin   = tin

A duet has two people. Caduet adds a Calcium channel blocker to another drug. C-a for calcium on the Periodic Table of Elements, d-u-e-t for duet. C-a-d-u-e-t, Caduet.

275. Amlodipine/ Benazepril (Lotrel) *(LOE-trel)*

am   = (l)am(p), triamcinolone
lo'   = (Co)lo(rado), (s)lo(w), loperamide
di   = di(nner), di(n), cefdinir

pine   = (shi)p(mar)ine, atropine

be   = be(d), rabeprazole
na'   = na(p), (g)na(t), phenazopyridine
ze   = ze(d), ze(phyr), azelastine
pril   = pril(l), benazepril

276. Amlodipine / Valsartan (Exforge) *(ECKS-forge)*

am   = (l)am(p)
lo'   = (Co)lo(rado)
di   = di(nner)
pine   = (shi)p(mar)ine

val   = val(ley)
sar'   = sar(dine)
tan   = tan

277. Felodipine (Plendil) *(PLEN-dill)*

fe   = fe(n)
lo'   = (Co)lo(rado)
di   = di(nner)
pine   = (shi)p(mar)ine

278. Nifedipine (Procardia) *(pro-CARD-e-uh)*

ni   = ni(ght)
fe'   = fe(n)
di   = di(nner)
pine   = (shi)p(mar)ine

Nifedipine is another dihydropyridine with the "dipine, d-i-p-i-n-e" stem.

Procardia takes the "p-r-o" from "promotes" and "c-a-r-d-i-a" from "cardiac" so you can remember the brand **Procardia** promotes cardiac health.

VI. Vasodilators

279. Hydralazine (Apresoline) *(ah-PREZ-oh-leen)*

hy   = hy(brid)
dra'   = dra(gon)
la   = (co)la
zine   = (maga)zine

See the "p-r-e-s, pres" from blood pressure in **Apresoline.**

280. Isosorbide mononitrate (Imdur) *(IM-dur)*

i'   = "i"
so   = so(fa)
sor'   = sor(bet)
bide = bide

mo'   = mo(nolith)
no   = no
ni'   = ni(ght)
trate = (concen)trate

281. Nitroglycerin (Nitrostat) *(NYE-trow-stat)*

ni   = ni(ght)
tro   = (me)tro
gly' = gly(ph)
ce   = ce(nt)
rin   = rin(se)

"Nitro-"is a World Health Organization (W-H-O) stem.

Nitroglycerin converts to nitric oxide, a vasodilator so make sure the patient sits when he takes the med because it causes significant dizziness.

With the brand **Nitrostat**, take the n-i-t-r-o from "nitrous" as in a street car's extra fuel and the concern that the patient and blood pressure drop quickly or "stat, s-t-a-t"; **Nitrostat.**

VII. Anti-anginal

282. Ranolazine (Ranexa) *(rah-NECKS-ah)*

ra   = (au)ra
no'   = no
la   = (co)la
zine = (maga)zine

Ranolazine (Ranexa) and **hydralazine (Apresoline)** both have similar endings, but **ranolazine** *treats* angina pectoris and **hydralazine** may *cause* angina pectoris. I grouped **hydralazine** with another vasodilator, **isosorbide dinitrate**. The "<u>nitrate</u>" helps you remember this is in the same class as **nitroglycerin**.

171

VIII. Antihyperlipidemics

<u>HMG-CoA reductase inhibitors</u>

283. Ator<u>vastatin</u> (Lipitor) *(LIP-eh-tore)*

a   = (comm)a
tor'   tor(so)
va   = (lar)va
sta   = sta(ff)
tin   = tin

Start with the infix "va, v-a" and add s-t-a-t-i-n to get "vastatin, v-a-s-t-a-t-i-n" as a way to identify the H-M-G-Co-As.

Students use letters of the H-M-G-Co-A class to memorize potential adverse effects: "H" for <u>h</u>epatotoxicity, "M" for <u>m</u>yositis, "G" for <u>g</u>estation (that you can't use atorvastatin during pregnancy).

Brand name **Lipitor** is a <u>lip</u>id gladia<u>tor</u>, taking l-i-p from lipid and t-o-r from gladiator; **Lipitor**.

284. Lo<u>vastatin</u> (Mevacor) *(MEV-uh-core)*

lo'   = (Co)lo(rado)
va   = (lar)va
sta   = sta(ff)
tin   = tin

I listed the HMG-CoA reductase inhibitors **lovastatin**, **pravastatin** and **simvastatin** alphabetically, each has the statin stem with v-a substem to make –vastatin. The "cor, c-o-r" in Mevacor hints at <u>cor</u>onary.

285. Pravastatin (Pravachol) *(PRAV-uh-coll)*

pra' = pra(nce)
va = (lar)va
sta = sta(ff)
tin = tin

The "chol" in **Pravachol** hints at cholesterol.

286. Rosuvastatin (Crestor) *(CRES-tore)*

ro = ro(w)
su' = su(e)
va = (lar)va
sta = sta(ff)
tin = tin

Like **atorvastatin, rosuvastatin** shares the "vastatin, v-a-s-t-a-t-i-n" ending. Remember **Crestor** decreases cholesterol, taking the c-r from decreases and e-s-t-o-r from cholesterol to make **Crestor**.

287. Simvastatin (Zocor) *(ZOE-core)*

sim' = sim(ple)
va = (lar)va
sta = sta(ff)
tin = tin

Fibric acid derivatives

288. Feno<u>fibrate</u> (TRICOR) *(TRY-core)*

fe	= fe(n)
no	= no
fi'	= fi(n)
brate	= (vi)brate

Alternate pronunciation:

fe	= fe(ed)
no	= no
fi'	= fi(ve)
brate	= (vi)brate

A drug like **fenofibrate** has the obvious stem "fibrate, f-i-b-r-a-t-e," a triglyceride lowering fibric acid derivative.

The brand name **TRICOR** takes the t-r-i from <u>tri</u>glycerides and c-o-r from <u>cor</u>onary; **TRICOR**.

289. Gem<u>fibro</u>zil (Lopid) *(LOE-pid)*

gem	= gem
fi'	= fi(n)
bro	= bro(ther)
zil	= (Bra)zil

Alternate pronunciation:

gem	= gem
fi'	= fi(ve)
bro	= bro(ther)
zil	= (Bra)zil

You can see the "fib" in their <u>fib</u>ric acid derivative **gem<u>fibro</u>zil.** That brand name **Lopid** hints at "<u>lo</u>wering li<u>pid</u>s."

Bile acid sequestrant

290. Colesevelam (Welchol) *(WELL-call)*

co   = co(ne)
le   = le(tter)
se'   = se(t)
ve   = ve(teran)
lam   = lam(b)

The bile acid sequestrant **colesevelam (Welchol)** has both a generic and brand name with a hint at cho̲l̲e̲sterol and the brand name at getting "well". Welchol.

Cholesterol absorption blocker

291. Ezetimibe (Zetia) *(ZEH-tee-ah)*

e	= (n)e(t)
ze'	= ze(d)
ti	= ti(c)
mibe	= m(y)(tr)ibe

is a cholesterol absorption blocker, a different type of drug, and can combine with an HMG-CoA like simvastatin (Zocor) to make Vytorin.

292. Ezetimibe / Simvastatin (Vytorin) *(vie-TOR-in)*

e	= (n)e(t)
ze'	= ze(d)
ti	= ti(c)
mibe	= m(y)(tr)ibe
sim'	= sim(ple)
va	= (lar)va
sta	= sta(ff)
tin	= tin

IX. Anticoagulants and antiplatelets

<u>Anticoagulants</u>

293. Enoxaparin (Lovenox) *(LOW-ven-ox)*

e	= (n)e(t)
nox	= nox(ious)
a	= (comm)a
pa'	= pa(ir)
rin	= rin(se)

Enoxaparin and **heparin** share the "parin, p-a-r-i-n" stem because they are related. **Enoxaparin** is more expensive per dose, but patients can use it at home. It's also used as bridge therapy in a patient who is starting **warfarin** therapy.

Lovenox is a <u>lo</u>w molecular weight heparin for deep <u>ven</u> thrombosis prevention, taking the l-o from low and v-e-n from vein, **Lovenox**.

294. **Heparin** *(HEP-uh-rin)*

he'	= he(n)
pa	= (pa)pa
rin	= rin(se)

Heparin and "bleedin'" sort of rhyme to remember its primary adverse effect. A student mentioned the actor Dennis Quaid's twins, who received a double dose of **heparin** that caused bleeding. Sometimes knowing a celebrity who has delt with a condition helps place the drug in memory.

<u>2</u>95. War<u>far</u>in (Coumadin) *(KOO-ma-din)*

war'   = war
fa   = so(fa)
rin   = rin(se)

That the "parin, p-a-r-i-n" stem from the anticoagulants **heparin** and **enoxaparin** and "farin, f-a-r-i-n" stem of **warfarin** are similar. This reminds students they are all anticoagulants.

Students associate bleeding with <u>warfare</u> and warfarin has the w-a-r-f-a-r from warfare to recognize bleeding is a potential adverse effect.

The I-N-R (<u>i</u>nternational <u>n</u>ormalized <u>r</u>atio), a way of measuring **warfarin's** effectiveness, monitors the patient who is on therapy.

"I-N-R" happen to be the last three letters of **warfa<u>rin</u>** to help you remember.

A student said that **war<u>far</u>in** has "far, f-a-r" in it, as in you have to go far to have blood drawn.

A way to remember Vitamin K affects warfarin's brand name **Coumadin** and coagulation is to spell **Coumadin** with a "K" instead of a "C." K-o-u-m-a-d-i-n.

296. Dabigatran (Pradaxa) *(pra-DAX-uh)*

da   = (so)da
bi'   = bi(t)
ga   = (yo)ga
tran   = tran(sfer)

Memorize **dabigatran's** "gatran, g-a-t-r-a-n" stem to note the difference between anticoagulants.

Dabigatran doesn't need INR monitoring like **warfarin** does and you can note the last three letters in **dabigatran** as not being "I-N-R."

Use the b-i, bi in dabigatran to remember it's a factor IIa [[two-"A"]] inhibitor.

297. Rivaroxaban (Xarelto) *(zah-REL-toe)*

ri   = ri(de)
va   = (lar)va
rox'   = (xe)rox
a   = (comm)a
ban   = ban(d)

Alternate pronunciation:

ri   = ri(nse)
va   = (lar)va
rox'   = (xe)rox
a   = (comm)a
ban   = ban(d)

has x-a in the name. This is important because its mechanism of action is to block facter Xa [[10-"a"]] expressed as Roman numeral "X" meaning 10 and "a". Blocking this factor blocks coagulation. Or, you can think of banning coagulation using the last three letters b-a-n, ban in **rivaroxaban**. The brand name Xarelto starts with a capital "X" and little "a" again pointing to factor Xa.

298. Api<u>xaban</u> (Eliquis) *(ELL-eh-kwis)*

a	= (comm)a
pix'	= pix(el)
a	= (comm)a
ban	= ban(d)

has the same x-a-b-a-n, -xaban stem as **rivaroxaban** and also blocks factor Xa [[10-A]].

The anticoagulants **dabigatran (Pradaxa), rivaroxaban (Xarelto),** and **Apixaban (Eliquis)** are usually coupled because they don't require monitoring like their counterpart, **warfarin (Coumadin).**

<u>Antiplatelet</u>

299. Aspirin / Dipyridamole (Aggrenox) *(AGG-reh-nocks)*

as'	= as(p)
pi	= pi(n)
rin	= rin(se)

di'	= di(ce)
py	= py(xis)
ri	= ri(nse)
da'	= (so)da
mole	= mole

work together to prevent clots and the brand name can be thought of as "<u>agg</u>regate <u>not</u>," **Aggrenox**.

300. Clopidogrel (Plavix) *(PLA-vix)*

clo	= clo(ver)
pi'	= pi(n)
do	= do(ugh)
grel	= (mon)grel

Clopidogrel has the "grel, g-r-e-l" stem identifying its class.

Clopidogrel and **aspirin** work similarly leading to a reduced likelihood that platelets will stick together and clot.

Plavix <u>vex</u>es <u>pla</u>telets taking the "v" and "x" from vexes and p-l-a from platelets to keep the blood thin; **Plavix**.

301. Prasugrel (Effient) *(EH-fee-ent)*

pra'	= pra(nce)
su	= su(e)
grel	= (mon)grel

Prasugrel shares the "–grel" stem with **clopidogrel**. **Effient** is the word "efficient" without the "c-i," so may be efficient at thinning platelets

302. Ticagrelor (Brilinta) *(Brih-LINN-tah)*

ti	= ti(e)
ca'	= ca(t)
gre	= gre(mlin)
lor	= lor(d)

Ticagrelor also has the "grel, g-r-e-l" stem identifying its class.

Ticagrelor has demonstrated superiority to **clopidogrel**.

X. Cardiac glycoside and Anticholinergic

<u>Cardiac glycoside</u>

303. Digoxin (Lanoxin) *(la-KNOCKS-in)*

di   = di(nner)
gox'   = (oxy)g(en)(b)ox
in   = in

Digoxin treats congestive heart failure by increasing the force of the heart's contractions.

Digoxin is derived from the plant *Digitalis lanata*. In Latin, *Digitalis* means something like hand or "digits," while *lanata* means "wooly" because the actual plant is fuzzy.

Therefore, **digoxin** takes the d-i-g from *digitalis*, and the brand name **Lanoxin** takes the l-a-n from *lanata*.

Alternatively, you could remember that **Lanoxin** and **digoxin** keep your heartbeat r<u>o</u>ck<u>in</u>'.

Anticholinergic

304. Atropine (AtroPen) *(ah-trow-PEN)*

a^l	= (m)a(t)
tro	= (me)tro
pine	= (shi)p(mar)ine

Atropine has the –tropine, t-r-o-p-i-n-e stem.

Atropine causes anticholinergic (anti = against, cholinergic = of acetylcholine) effects.

Anticholinergic effects fall under the broad category of "dry."

Use the ABDUCT, capital A-B-D-U-C-T mnemonic, as in anticholinergics "abduct" water to remember these six effects.

Anhidrosis (A), Blurry vision (B) (secondary to dry eyes), Dry mouth (D), Urinary retention (U), Constipation (C), and Tachycardia (T).

This tachycardic side effect therapeutically prevents bradycardia in patients undergoing certain procedures.

Note the opposite: Cholinergic effects would include "wet" effects: sweating, lacrimation (watery eyes), hypersalivation, urinary incontinence, diarrhea, and bradycardia.

XI. Antidysrhythmic

305. Amiodarone (Cordarone) *(CORE-dah-rone)*

a   = (m)a(t)
mi'   = (se)mi [Note: rhymes with me]
o   = "o"
da   = (so)da
rone = (d)rone

The generic name **Amiodarone** and brand name **Cordarone** share the "arone" lettering. Cardiologists can also use beta-blockers, calcium channel blockers, and **digoxin** as antidysrhythmics.

CHAPTER 7 - ENDOCRINE / MISC.

I. OTC Insulin and emergency contraception

306. Regular insulin (Humulin R) *(HUE-myou-lin ARE)*

in'   = in
su   = su(b)
lin   = lin(t)

Regular insulin is short acting. Don't confuse this insulin with the shortest-acting insulins available, such as insulin lispro, brand **Humalog**. Prescribers give **regular insulin** when patients need to adjust dosages on a sliding scale.

Insulin used to come from a pig (porcine) or cow (bovine), but now matches human insulin because of molecular engineering.

Therefore, the Eli Lilly brand name **Humulin** simply squishes the words <u>hum</u>an and ins<u>ulin</u> together, taking the h-u-m from human and u-l-i-n from insulin; **Humulin**.

307. NPH insulin (Humulin N) *(HUE-myou-lin EN)*

in'   = in
su   = su(b)
lin   = lin(t)

The "N-P-H" in **NPH insulin** stands for <u>n</u>eutral <u>p</u>rotamine <u>H</u>agedorn. The "neutral protamine" refers to how Hagedorn, the inventor, chemically altered the insulin.

The "e-n" pronunciation of the letter "N" sounds a little like "i-n" and can help you remember it's an <u>in</u>termediate acting <u>in</u>sulin, taking the i-n from both intermediate and insulin; **Humulin N**.

308. Levonorgestrel (Plan B One-Step) *(plan-bee won-step)*

le	= le(tter)
vo	= vo(te)
nor	= nor
ges'	= ges(t)
trel	= trel(lis)

You can recognize **levonorgestrel** as a progestin hormone product by the "gest, g-e-s-t" stem. Take it within 72 hours after sexual intercourse. It causes nausea, but flat soda can calm this down. It's now called **Plan B One-Step** because it used to take two steps or two doses to provide this contraception.

I used to work in a college town pharmacy. Every Saturday and Sunday morning, I would have many students come to pick up **Plan B**. Every time, the man drove, and the woman sat in the passenger seat. When I told the male student it was fifty bucks for the Plan B, he would invariably look at her, and then pay. One time, however, I overheard her say, "Oh no, you *just* didn't." from that "just, j-u-s-t" remember the "gest, g-e-s-t" that is part of **levonorgestrel**.

II. Diabetes and insulin

<u>Biguanides</u>

309. Met<u>formin</u> (Glucophage) *(GLUE-co-fage)*

met   = (co)met
for'   = for
min   = min(t)

Use the "formin, f-o-r-m-i-n" stem to remember **metformin** is a biguanide anti-diabetic.

One student came up with this mnemonic, "If you <u>met four</u> <u>men</u> on **Glucophage**, they are diabetic then."

Phagocytosis is the process of cell eating. You can use the brand name **Glucophage,** G-l-u-c-o-p-h-a-g-e to think of the medication as eating, taking the g-l-u-c-o from <u>gluco</u>se and p-h-a-g-e from <u>phage</u>-ing; **Glucophage**.

310. Met<u>formin</u> / Gl<u>y</u>buride (Glucovance) *(GLUE-ko-vance)*

met   = (co)met
for'   = for
min   = min(t)

gly'   = gly(cemic)
bu   = bu(reau)
ride   = ride

I alphabetized the antidiabetic oral medications by class. Manufacturers combine the biguanide **metformin (Glucophage)** with the sulfonylurea **glyburide (DiaBeta)** to make **Glucovance** – an ad<u>vance</u> in <u>gluco</u>se lowering.

DPP-4 Inhibitors (Gliptins)

Patients refer to DPP-4 inhibitors **linagliptin** and **saxagliptin** by their stem "-gliptin."

311. Linagliptin (Tradjenta) *(trah-GEN-tah)*

li	= li(ly)
na	= (bana)na
glip'	= (bi)g(s)lip
tin	= tin

312. **Saxagliptin (Onglyza)** *(on-GLY-zah)*

sax	= sax(ophone)
a	= (comm)a
glip'	= (bi)g(s)lip
tin	= tin

313. **Sitagliptin (Januvia)** *(ja-NEW-vee-uh)*

si	= si(t)
ta	= (da)ta
glip'	= (bi)g(s)lip
tin	= tin

Although the "gliptin, g-l-i-p-t-i-n" stem helps us recognize **sitagliptin** as an anti-diabetic, students associate the sugar you might put in "Lipton, L-i-p-t-o-n" iced tea with **sitagliptin**.

The brand name **Januvia** ends in "via, v-i-a" and is similar to the "dia, d-i-a" from diabetes.

Meglitinides (Glinides)

We refer to the the meglitinide **repaglinide**, as a "-glinide," again based on its suffix.

314. Repaglinide (Prandin) *(PRAN-din)*

re   = re(d)
pa'   = pa(th)
gli   = gli(b)
nide   = (s)nide

The brand name **Prandin** includes the part of the word "prandial" which relates to dinner or lunch, something diabetics are always aware of as they nee to watch carbohydrate content.

Sulfonylureas – 2ⁿᵈ-generation

315. Glipizide (Glucotrol) *(GLUE-co-trawl)*

gly' = gly(cemic)
bu   = bu(reau)
ride = ride

Your first question should be, "what happened to the first generation sulfonylureas?" Sometimes, when a medication's next generation so surpasses the previous as the preferred product, we simply don't include them anymore. First generation sulfonylureas cause more effects that are adverse. There is little reason to use them anymore.

The "gli-, g-l-i" stem in **glipizide** indicates an antihyperglycemic medication. The brand name **Glucotrol** alludes to the <u>glucose</u> con<u>trol</u> in diabetics taking the g-l-u-c from glucose and t-r-o-l from control; **Glucotrol**.

316. Glimepiride (Amaryl) *(AM-ah-ril)*

gli   = gli(b)
me'   = me(n)
pi   = pi(n)
ride = ride

Like **glipizide**, **glimepiride** contains the g-l-i, gli stem.

317. Glyburide (DiaBeta) *(die-uh-BAY-ta)*

gli'   = gli(b)
pi   = pi(n)
zide = z(ipper)(r)ide

The "gly, g-l-y" stem in **glyburide** is archaic. The stem "gli, g-l-i," replaces it in new sulfonylurea antidiabetics.

The brand name **DiaBeta** combines the "dia, d-i-a" from diabetic, and the "B-e-t-a" from the Beta cells, which release insulin; **DiaBeta**.

While diabetes is an issue with excess glucose, hypoglycemia, too little glucose, also requires treatment.

Thiazolidinediones (Glitazones)

Like the gliptins and glinides we refer to the thiazolidinediones **pioglitazone** and **rosiglitazone** as "-glitazones" using their stems to make it easier.

318. Pioglitazone (Actos) *(ACK-toes)*

pi   = pi(e)
o   = "o"
gli'   = gli(b)
ta   = (da)ta
zone = zone

One student said that if you eat too much pie, you'll need pie-o-glitazone for your diabetes.

319. Rosiglitazone (Avandia) *(ah-VAN-dee-ah)*

ro   = ro(w)
si   = si(t)
gli'   = gli(b)
ta   = (da)ta
zone = zone

Incretin mimetics

320. Exenatide (Byetta) (bye-YET-ah)

ex   = ex(po) [Note: not "egg zit"]
e'   = (n)e(t)
na   = (bana)na
tide   = tide

321. Liraglutide (Victoza) *(vick-TOE-zah)*

li   = li(ly)
ra'   = ra(ttle)
glu   = glu(ten)
tide   = tide

Both incretin mimetics **exenatide (Byetta)** and **liraglutide (Victoza)** are injectables.

Hypoglycemia

322. Glucagon (GlucaGen) *(glue-ca-JEN)*

glu' = glu(ten)
ca = (or)ca
gon = (hexa)gon

Remember the generic name glucagon with, "I use **glucagon** when the glucose is gone," taking the g-l-u-c from glucose and g-o-n from gone; **Glucagon**.

GlucaGen, the brand name, is a glucose generator taking the g-l-u-c from glucose and g-e-n from generator; **GlucaGen**.

<u>RX Insulin</u>

323. Insulin aspart (Novolog) *(NOE-voh-log)*

in'	= in
su	= su(b)
lin	= lin(t)
as'	= as(p)
part	= part

is another insulin analog.

324. Insulin lispro (Humalog) *(HUE-mah-log)*

in'	= in
su	= su(b)
lin	= lin(t)
lis'	= lis(t)
pro	= pro(ton)

Insulin glargine would precede **insulin lispro** alphabetically, but the convention is to list the medications shorter acting to longer acting. The complete order from shortest to longest acting is 1) insulin lispro, 2) regular insulin, 3) insulin NPH, and 4) insulin glargine.

I use my Spanish language to help memorize that **insulin lispro** is rapid acting. When I was in Mexico on a zip line, the person in the first tower would say, "Listo, listo," [[lee-stoh, lee-stoh]] meaning "You ready, I'm ready." Then I would fly fast down that zip cord. I just replace the lispro's "s-p," with "t" because you must be "listo, listo" to take this rapid acting insulin with a meal.

Insulin lispro's brand name, **Humalog,** is a <u>huma</u>n insulin ana<u>log</u> combining the h-u-m from human and l-o-g from analog. I always pictured a <u>huma</u>n on a <u>log</u> floating down a rapids river to remember **Humalog** as a rapid acting insulin.

325. Insulin detemir (Levemir) *(LEH-veh-meer)*

in'   = in
su   = su(b)
lin   = lin(t)

de'   = de(sk)
te   = te(nt)
mir   = mir(ror)

Alternate pronunciation:

de'   = de(sk)
te   = te(nt)
mir   = (ad)mir(al)

Insulin detemir is a long-acting insulin like **Lantus**.

326. Insulin glargine (brand names Lantus and Toujeo)
(LAN-tuss, TWO-jzeh-oh)

in'   = in
su   = su(b)
lin   = lin(t)

glar'   = g(old)lar(k)
gine   = (aspara)gine

With **insulin glargine**, I think of the g-l-a-r in gla_ring_ and i-n of lurking, someone who is *slowly* creeping around.

For the brand **Lantus**, one student came up with "**Lantus** la_sts_ all day long, take it at night, and your life will be prolonged." I've also heard "_la_zy **Lantus**" used to help remember it's a 24-hour drug.

III. Thyroid hormones

<u>Hypothyroidism</u>

327. Levothyroxine (Synthroid) *(SIN-throyd)*

le	= le(an)
vo	= vo(te)
thy	= thy(roid)
rox'	= (xe)rox
ine	= (fam)ine

Alternate pronunciation:

le	= le(tter)
vo	= vo(te)
thy	= thy(roid)
rox'	= (xe)rox
ine	= (sal)ine

The generic name **levothyroxine** has the "thyro, t-h-y-r-o" from <u>thyro</u>id in the name. You just have to remember it's for supplementation.

The brand name **Synthroid** combines the words <u>synth</u>etic and thy<u>roid</u>, taking the s-y-n-t-h from synthetic and r-o-i-d from thyroid; **Synthroid**.

Hyperthyroidism

328. Propylthiouracil (P-T-U) *(pee-tee-you)*

pro   = pro(ton)
pyl   = (po)p(vin)yl
thi   = thi(gh)
o   = "o"
u'   = "u"
ra   = (au)ra
cil   = (pen)cil

PTU takes "p" from propyl, "t" from "thio," and "u" from uracil in the generic **propylthiouracil**.

Although "thio, t-h-i-o" suggests a sulfur atom, you can think of it as thyroid lowering, using t-h-i from thyroid and "o" from lowering; propyl*thio*uracil.

IV. Hormones and contraception

<u>Low testosterone</u>

329. Test<u>oste</u>rone (AndroGel) *(ANN-droh GEL)*

tes = tes(t)
to' = to(t)
ste = ste(ps)
rone = (d)rone

Most people know the steroid hormone **testosterone**, but note the stem for a steroid is "ster, s-t-e-r" "<u>Andro</u>, A-n-d-r-o" is the Greek prefix for male and <u>gel</u> is the vehicle in **AndroGel**, indicating it's a "male's gel."

Estrogens and / or Progestins

In this section, I list estrogens first, then combo estrogen /
progestin products, then progesterone products, recognizable
by their "estr-" and "-gest-" stems respectively.

330. Estradiol (Estrace, Estraderm) *(ES-trace, ES-tra-derm)*

es = (m)es(s)
tra = (orches)tra
di' = di(ce)
ol = (aw)ol

Generic **estradiol** and brand names **Estrace** and **Estraderm** all
have the e-s-t-r, estrogen stem.

331. Conjugated estrogens (Premarin) *(PREM-uh-rin)*

es' = (m)es(s)
tro = (me)tro
gens = (oxy)gens

Premarin is for "pregnant mare's urine," the drug's source.

332. Conjugated <u>estr</u>ogens / Medroxyprogesterone (Prempro, Premphase) *(PREM-prow, PREM-phase)*

es'	= (m)es(s)
tro	= (me)tro
gens	= (oxy)gens
me	= me(n)
drox	= dr(ink)(b)ox
y	= (countr)y
pro	= pro(ton)
ges'	= ges(t)
te	= te(nt)
rone	= (d)rone

Prempro and Premphase have different estrogen / progestin levels.

333. Pro<u>ges</u>terone (Prometrium) *(pro-MEH-tree-uhm)*

pro	= pro(ton)
ges'	= ges(t)
te	= te(nt)
rone	= (d)rone

334. Medroxypro<u>ges</u>terone (Provera) *(pro-VER-ah)*

me	= me(n)
drox	= dr(ink)(b)ox
y	= (countr)y
pro	= pro(ton)
ges'	= ges(t)
te	= te(nt)
rone	= (d)rone

Progesterone and medroxyprogesterone are both progestin tablets.

Combined oral contraceptive pill (COCP)

335. Norethindrone with ethinyl estradiol and ferrous fumarate (Loestrin 24 Fe [[ef-ee]]) *(Low-ES-trin EF-ee Twen-TEE fore)*

eth'	= (m)eth(od)
i	= i(t)
nyl	= (vi)nyl
es	= (m)es(s)
tra	= (orches)tra
di'	= di(ce)
ol	= (aw)ol
nor	= nor
eth'	= (m)eth(od)
in	= in
drone	= drone

It's important to first memorize the estrogen stem "estr, e-s-t-r" and progestin stem "gest, g-e-s-t."

To remember the brand, use this mnemonic: **Loestrin 24 Fe** reduces the length of menstruation, with iron supplementation, to prevent an anemic situation."

336. Norgestimate with ethinyl estradiol (Tri-Sprintec) *(tri-SPRIN-tech)*

eth'	= (m)eth(od)
i	= i(t)
nyl	= (vi)nyl
es	= (m)es(s)
tra	= (orches)tra
di'	= di(ce)
ol	= (aw)ol
nor	= nor
ges'	= ges(t)
ti	= ti(c)
mate	= mate

Again, look to the estrogen stem "estr, e-s-t-r" and progestin stem "gest, g-e-s-t."

Here's another mnemonic: **"Tri-Sprintec is triphasic, take three different doses, in seven-day spaces."**

Patch

337. Norelgestromin with ethinyl estradiol (OrthoEvra)
(OR-thoe EV-rah)

eth'   = (m)eth(od)
i   = i(t)
nyl   = (vi)nyl

es   = (m)es(s)
tra   = (orches)tra
di'   = di(ce)
ol   = (aw)ol

nor   = nor
el   = el(bow)
ges' = ges(t)
tro   = (me)tro
min = min(t)

One student associated **norelgestromin's** "Norel" as "not oral" to remember this is a patch. Another mnemonic highlights the drug's correct placement and length of therapy.

"OrthoEvra is a patch, put it on your arm, your abs, your buttock or back, and then take it off a week after that."

Ring

338. Etonogestrel with ethinyl estradiol (NuvaRing) *(NEW-va-ring)*

eth'	= (m)eth(od)
i	= i(t)
nyl	= (vi)nyl
es	= (m)es(s)
tra	= (orches)tra
di'	= di(ce)
ol	= (aw)ol
e	= (n)e(t)
to	= (pis)to(n)
no	= no
ges'	= ges(t)
trel	= trel(lis)

A student thought of **etonogestrel**'s "Etono" in as "Eat, oh no" to remember it's not an oral tablet.

The brand **NuvaRing** combines "n-u" from new, "v-a" from vaginal, and "r-i-n-g" from "ring;" **Nuvaring**.

V. Overactive bladder, urinary retention, erectile dysfunction (ED), benign prostatic hyperplasia (BPH)

<u>Overactive bladder</u>

339. Oxybutynin (brand names Ditropan and Oxytrol O-T-C) *(DIH-trow-pan OX-ee-trawl OTC)*

ox	= ox
y	= (countr)y
bu'	= bu(reau)
ty	= ty(pical)
nin	= (mela)nin

One student remembered **oxybutynin** as keepin' the urin' in.

Both **Di*tro*pan** and **Oxy*tro*l** have the "t-r-o" from control for contro<u>lli</u>ng an overactive bladder.

340. Dari<u>fe</u>nacin (Enablex) *(EN-ah-blecks)*

da	= da(rt)
ri	= ri(nse)
fe'	= fe(n)
na	= (bana)na
cin	= cin(der)

Dari<u>fe</u>nacin and **soli<u>fe</u>nacin** ave the same "–fenacin" stem. Both work for overactive bladder (OAB). The **Enablex** brand name hints at "<u>enabling</u>" the patient to "<u>exit</u>" the house when the OAB might have kept them in.

341. Soli<u>fe</u>nacin (VESIcare) *(VEH-si-care)*

so	= so(fa)
li	= li(ly)
fe'	= fe(n)
na	= (bana)na
cin	= cin(der)

Solifenacin's stem, "fenacin, f-e-n-a-c-i-n," should be your first clue to its function.

We give **solifenacin** once daily, so thinking about it as "slow-fenacin" helps in remembering this point.

In addition, **solifenacin** solves the problem of urine that needs to be "<u>fenc</u>ed <u>in</u>," using the "f-e-n-c" from fenced and "i-n" to almost complete the stem "fenacin."

The brand **VESIcare** contains *vesica*, v-e-s-i-c-a, which means "bladder" in Latin, **VESIcare**.

342. Tolterodine (Detrol) *(DEH-trawl)*

tol	= tol(erate)
te'	= te(nt)
ro	= ro(w)
dine	= (bir)di(e)(tu)ne

The generic name **tolterodine** has the "t-r-o" from con<u>tro</u>l in the name as well.

The brand **Detrol** helps <u>det</u>rusor muscle con<u>trol</u>, keeping urine in, taking the d-e-t-r from detrusor and –r-o-l from control; **Detrol**.

Urinary retention

343. Bethanechol (Urecholine) *(yur-eh-CO-lean)*

be   = be(d)
tha'   = tha(tch)
ne   = ne(t)
chol   = (s)chol(ar)

The "chol, c-h-o-l" in **bethane<u>chol</u>** helps you remember it's *chol*inergic.

While *anti*cholinergics are dry, cholinergics do the opposite and make things wet.

This drug assists the bladder muscles in expelling urine for a patient with urinary retention.

The brand name **Urecholine** alludes to how it affects <u>ur</u>ination through <u>chol</u>inergic effects, taking the u-r from urination and c-h-o-l from cholinergic; **Urecholine**.

Erectile dysfunction - PDE-5 inhibitors

344. Sildenafil (Viagra) *(vie-AG-rah)*

sil	= sil(ver)
de'	= de(sk)
na	= (bana)na
fil	= fil(m)

The –afil, a-f-i-l stem specifies the P-D-E-5 inhibitor class.

There is a scene with Jack Nicholson in the movie *Something's Gotta Give* that reminds us that patients shouldn't combine **sildenafil** with nitrates like **nitroglycerin** or they will end up in the emergency room.

The brand **Viagra** brings viable growth, taking the v-i-a from viable and –g-r from growth, – an erection.

345. Vardenafil (Levitra) *(leh-VEE-trah)*

var	= var(sity)
de'	= de(sk)
na	= (bana)na
fil	= fil(m)

Vardenafil seems to have the same "-den-" substem as **sildenafil**. To levitate is to rise above the ground, so the brand name **Levitra** hints at the rising erection.

346. Tadal<u>afil</u> (Cialis) *(see-AL-is)*

ta = (da)ta
da' = da(nce)
la = (co)la
fil = fil(m)

Again, the "afil, a-f-i-l" stem specifies the P-D-E-5 inhibitor class. **Cialis** is the weekend pill because it, unlike **sildenafil**, lasts the weekend with a long half-life.

I asked some students how they remembered **tadalafil**. I'm hesitant to share their mnemonic.

One said you just think "<u>ta</u>-<u>dah</u>" as in "surprise," to remember the first two syllables "tada, t-a-d-a."

I stopped them before they started on to how they remember the "fil" part of the generic name.

Cialis is the dual bathtub commercials drug.

BPH – Alpha-blocker

347. Alfuzosin (Uroxatral) *(YUR-ox-uh-trall)*

al	= (s)al(iva)
fu'	= fu(mes)
zo	= zo(mbie)
sin	= (ba)sin

Again, the "osin, o-s-i-n" ending is not an actual stem, but a way to connect **tamsulosin** and **alfuzosin**.

With BPH, I think the brand **Uroxatral** sounds a little like "Urine control," taking the u-r from urine and "t-r-o-l, t-r-o-l" sound from control to match Uroxatral's t-r-a-l. **Uroxatral.**

348. Tamsulosin (Flomax) *(FLOW-Max)*

tam	= tam(bourine)
su'	= su(e)
lo	= (Co)lo(rado)
sin	= (ba)sin

Flomax allows for a flow of maximum urine taking the f-l-o from flow and m-a-x from maximum; **Flomax.**

The "osin" ending is not an actual stem, but a way to connect **tamsul*osin*** and **alfuz*osin*** as similar.

BPH – 5-alpha-reducase inhibitor

349. Dutasteride (Avodart) *(AH-vo-dart)*

du   = du(et)
ta'   = ta(b)
ste   = ste(ps)
ride = ride

Use the "–steride, s-t-e-r-i-d-e" stem to recognize the 5-alpha-reductase inhibitors like **dutasteride**. The "ster, s-t-e-r" for steroid helps you remember it's for men (and prostate).

350. Finasteride (brand names Proscar and Propecia) *(PRO-scar, pro-PEE-shuh)*

fi   = fi(n)
na'   = na(p)
ste   = ste(ps)
ride = ride

Also employ the "–steride, s-t-e-r-i-d-e" stem to recognize the 5-alpha-reductase inhibitor **finasteride**

To connect the brand name **Proscar** to the generic **finasteride**, a professor told me of a student who used the phrase "That pro's car is the finest ride;" **Pros-car, finasteride**

The brand **Proscar** is for prostate care, taking the p-r-o from prostate and c-a-r from care. **Finasteride's** other brand name, **Propecia**, alludes to hair growth and is to reverse of alopecia (hair loss); pro-pecia (hair gain).

CHAPTER 8 - 30 NEWER DRUGS

It's tough to know for the board exams how up to date the actual exam is. For test makers, building the test is not just a matter of adding content, test questions need vetting and that takes many repititions. These newer drugs may or may not be on the current board examinations.

RESPIRATORY

1. Aclidinium bromide (Tudorza Pressair): Anticholinergic COPD medication. The c-l-i-d-i-n-i-u-m, clidinium stem means muscarinic antagonist or anticholinergic.

2. Umeclidinium bromide / vilanterol (Anoro Ellipta): Combination long-acting anticholinergic and ultra-long-acting Beta$_2$ agonist for COPD. The c-l-i-d-i-n-i-u-m, clidinium stem means muscarinic antagonist or anticholinergic and the t-e-r-o-l, terol stem is for beta-2 receptor agonist. The big advantage here is that vilanterol is once daily dosing.

3. Fluticasone furoate / vilanterol (Breo Ellipta): Inhaled corticosteroid and ultra-long-acting Beta$_2$ agonist for COPD combining the antiinflammatory fluticasone with once daily vilanterol.

4. Indacaterol (Arcapta Neohaler): Ultra-long-acting beta receptor agonist dosed once daily with the terol, t-e-r-o-l stem.

5. Roflumilast (Daliresp): Long acting phosphodiesterase-4 (PDE-4) inhibitor for COPD with the m-i-l-a-s-t, milast stem.

The d-a-l-i is pronounced as daily dosing for r-e-s-p, respiratory function.

IMMUNE

While all the Hepatitis C antivirals have v-i-r, vir in them there are more specific stems: previr, p-r-e-v-i-r, asvir, -a-s-v-i-r, buvir, b-u-v-i-r, and navir, n-a-v-i-r further subclassify these antivirals.

6. Bocep<u>revir</u> (Victrelis): Protease inhibitor antiviral for Hepatitis C.

7. Ledip<u>asvir</u> / sofos<u>buvir</u> (Harvoni): Two drug combination for Hepatitis C.

8. Ombit<u>asvir</u> / parita<u>previr</u> / rito<u>navir</u> / dasa<u>buvir</u> (Viekira Pak): An antiviral quad combination for Hepatitis C.

9. Simep<u>revir</u> (Olysio): Treatment for Hepatitis C.

10. Sofos<u>buvir</u> (Sovaldi): Treatment for Hepatits C.

11. Telap<u>revir</u> (Incivek and Incivo): Protease inhibitor antiviral for Hepatitis C.

NEURO

12. Brexpip<u>razole</u> (Rexulti): Atypical antipsychotic drug that is a D_2 partial agonist, a serotonin-dopamine activity modulator (SDAM) for schizophrenia. As with aripiprazole, watch out for brexpiprazole, which has the -prazole stem from proton pump inhibitors inside its own -piprazole stem. As Brexit was Britain's break from the U.K. so too is brex-piprazole for schizophrenia with schism meaning break.

13. Cariprazine (Vraylar): Atypical antipsychotic that works as a partial agonist on D2 and D3 receptors, with D3 selectivity for schizophrenia.

14. Suvorexant (Belsomra): Sedative / hypnotic that works as a selective, dual orexin receptor antagonist.

CARDIOVASCULAR

15. Aliskiren (Tekturna): Antihypertensive, direct renin inhibitor.

16. Alirocumab (Praluent): Proprotein convertase subtilisin / kexin type 9 (PCSK9) inhibitor for cholesterol.

17. Evolocumab (Repatha: Proprotein convertase subtilisin / kexin type 9 (PCSK9) inhibitor for cholesterol.

Alirokumab and evolocumab are currently the newest treatment options for high cholesterol. These agents are often for people who have a genetic disorder that leads to very high levels of cholesterol (called Familial Hypercholesterolemia). They are very expensive, but also very effective, lowering cholesterol levels more than even the highest doses of –statins.

18. Idarucizumab (Praxbind): Reverses anticoagulant effects of dabigatran (Pradaxa).

Again, you may hear idar-u-ciz-u-mab with a pronunciation of the letter "u" both times, but this convolutes the two important infixes, c-i, and z-u, which represent a cardiovascular disease class infix and a humanized source infix respectively.

19. Sacubitril / valsartan (Entresto):
Neprilysin inhibitor / Angiotensin II receptor blocker (ARB)

ENDOCRINE - WEIGHT LOSS

20. Lorcaserin (Belviq): Serotonergic weight loss drug, I think they were going for a brand name that rhymed with physique as in Belviq for your physique.

21. Phentermine and topiramate (Qsymia): Combination weight loss medication.

22. Pyridoxine/doxylamine (Diclegis): Combination of vitamin B6 and doxylamine succinate for morning sickness.

ENDOCRINE - DIABETES

23. Inhalable insulin (Afrezza): Inhalable insulin.

Inhalable insulin is an old drug (insulin) in a novel inhaled dosage form giving diabetics another treatment option. It is a rapid acting insulin like insulin lispro and aspart. Inhalable insulin's use is in smokers or those with asthma or COPD.

24. Insulin glargine (Toujeo): Long-acting once daily basal insulin analog.

Toujeo has the same active ingredient as Lantus, but Toujeo is 3 times as potent. You would think that a patient switching from Lantus to Toujeo would need a lower dose, but actually, patients that switch usually need about the same dose.

25. Insulin degludec (Degludec): Ultralong-acting basal insulin analog.

26. Canagliflozin (Invokana): Gliflozin class antidiabetic.

27. Dapagliflozin (Farxiga): Gliflozin class antidiabetic.

28. Empagliflozin (Jardiance): Gliflozin class antidiabetic.

Canagliflozin, dapagliflozin, and empagliflozin all share the stem –gliflozin, indicating that these medications are SGLT-2 inhibitors. SGLT-2 is a transporter within the kidneys that reabsorbs glucose. Inhibiting the transporter allows patients to urinate the suger lowering glucose levels. This leads to common UTIs as sugar is generally absent from the urine.

29. Albiglutide (Tanzeum): Glucagon-like peptide agonist (GLP-1 agonist)

30. Exenatide (Bydureon): Glucagon-like peptide agonist (GLP-1 agonist)

GLP-1 is a normal body hormone that increases insulin secretion, decreases glucagon secretion and decreases hunger. These are expensive medicines that need a subcutaneous administration.

INDEX

About the Author

Tony Guerra graduated with his Doctor of Pharmacy from the University of Maryland in Baltimore, Maryland and his Bachelor of Arts in English from Iowa State University and has been a practicing pharmacist for 19 years.

He lives in Ankeny, Iowa, with his wife Mindy and triplet daughters Brielle, Rianne, and Teagan.

Made in the USA
Middletown, DE
18 April 2021